I0050354

Competent Person duties?
(pg21)

First Aid training?
(pg74)

Postings at the job site?
(pg5)

Code of Safe Practices?
(pg21)

Heat stress training?
(pg86)

Crane operator certification?
(pg31)

Fire extinguishers on the job?
(pg73)

Heavy construction equipment?
(pg93)

Air compressor operating permits?
(pg12)

Qualified Person?
(pg122)

Fall protection systems?
(pg67)

Flagger training?
(pg75)

# Cal/OSHA Compliance FAQs

Hard hat requirements?
(pg117)

Supervisor training?
(pg161)

Multi-employer worksite rules?
(pg115)

Soil classification analysis?
(pg58)

Scaffold Permits?
(pg130)

Lockout/Blockout procedures?
(pg113)

Nighttime highway work zone illumination?
(pg112)

Permissible exposure limits for lead?
(pg108)

Flammable and combustible liquid storage?
(pg76)

Forklift operating rules?
(pg77)

Fall protection controlled access zone?
(pg71)

Atmospheric testing for confined space work?
(pg29)

# Construction Supervisor

# Cal/OSHA Compliance Guide

- Industrial
- Residential
- Commercial
- Heavy Civil

## Mike Leuck

www.OshaTools.com

Construction Supervisor
Cal/OSHA Compliance Guide

Copyright © 2013 by Mike Leuck

ISBN 978-1-935914-32-7

Printed in the United States of America

Order additional copies online at:

www.oshatools.com

Mike Leuck
Tailgate Publications
mikeleuck.csp@gmail.com
(831) 212-0093

Distributed by...

RIVER SANCTUARY PUBLISHING
P.O. Box 1561
Felton, California 95018
www.riversanctuarypublishing.com

## Acknowledgments

Cal/OSHA has an expert staff of Occupational Safety and Health Professionals working in Research and Education. Safety materials developed by this department were used as a resource in the writing of this Compliance Guide.

Other resources include: Occupational Safety and Health Administration (OSHA), National Institute for Occupational Safety and Health (NIOSH), Mine Safety and Health Administration (MSHA), Federal Highway Administration (FHWA), Federal Motor Carrier Safety Administration (FMCSA).

Each Agency has a website with access to their specific safety materials. Links to each Agency website are available at www.oshatools.com

## Disclaimer

The Cal/OSHA publication, "Cal/OSHA Pocket Guide for the Construction Industry," updated through June 2013, was referenced in the writing of this Guide. For the most up-to-date information, please refer directly to Title 8, California Code of Regulations.

This compliance guide is not all inclusive. The contents under the bullets, lists, notes, and exceptions are highlights of regulatory requirements, best practices and other construction safety and health information.

Readers shall refer directly to Title 8 of the California Code of Regulations and the Labor Code for detailed information regarding the regulation's scope, specifications, and exceptions and for other requirements that may be applicable to their operations.

This Guide refers the reader to various publications for additional information. Some of these publications are written by government agencies other than Cal/OSHA and might not follow Cal/OSHA rules exactly. The reader should be aware of this and use these resources accordingly.

This Guide is intended to be used as a safety compliance tool by Construction Supervisors. It should not be the only resource used. When in doubt about a compliance issue, the reader should always ask for assistance from professionals knowledgeable in the topic.

The author shall have neither liability nor responsibility to any person or company with respect to any loss or damage caused or alleged to be caused directly or indirectly by the information contained in this Guide.

# Contents

# Introduction

We were on break at our annual safety kick-off meeting. The speaker had just finished his motivational message of "watch out for the other guy's safety." I was listening in as a Construction Superintendent was talking with our CEO. "That was a good message," he said, "but what my Supervisors really need is the meat and potatoes of safety compliance. Supervisors struggle with knowing all that they need to know when it comes to safety."

That was the beginning of the "Construction Supervisor Cal/OSHA Compliance Guide."

The California construction industry involves many types of work activities covered by numerous regulations in Title 8 of the California Code of Regulations (T8 CCR).

T8 CCR contains detailed information on regulations and workplace safety programs including specifications and exceptions.

Depending on the work, Supervisors may need to know the Construction Safety Orders (CSOs), Electrical Safety Orders (ESOs), Tunnel Safety Orders (TSOs), Compressed Air Safety Orders (CASOs) and the General Industry Safety Orders (GISOs).

This Compliance Guide is intended to be a field reference for Supervisors and anyone else with safety responsibilities at a construction job. The Guide:

» Summarizes selected safety requirements from T8 CCR that apply to the construction industry.

» Provides highlights of selected safety standards in each major subject heading within its scope and may also include best practices in safety and health.

» It is not all inclusive and is not meant to be either a substitute for or a legal interpretation of the occupational safety and health regulations in Title 8 of the California Code of Regulations.

## How To Use This Guide

» Topics are listed alphabetically.

» Read from the beginning of each section for the best understanding of the topic.

» To read the complete Cal/OSHA T8 regulation on-line, "cut and paste" the regulation reference number listed with the topic into the T8 Regulation search engine, accessible through the T8 Search link at www.oshatools.com.

» To read "**Toolbox**" publications listed with a topic, "cut and paste" the **PUB###** (no spaces) listed with the publication into the search box at www.oshatools.com.

» **BONUS** - www.oshatools.com makes it easy for Supervisors to share safety info with the entire crew. Select the SHARE button located with each publication at www.oshatools.com, add the address, comments and send. No registration and no fee - just easy!

» Safety publications are continually being reviewed and added to www.oshatools.com. If you can't find a topic in this Compliance Guide, check www.oshatools.com. The topic you are looking for may have been added to the website.

# Access

The employer must provide safe access to and from all work levels or surfaces. Regulated means of access are as follows:

A. Stairways, ramps, or ladders must be provided at all points where a break in elevation of 18 in. or more occurs in a frequently traveled passageway, entry, or exit. **1629(a)(3)**

B. Aerial devices, such as cherry pickers and boom trucks, may be vehicle-mounted or self-propelled and used to position employees, tools, and materials. **3637, 3648**

C. Elevating work platforms, such as vertical towers and scissor lifts, are designed to raise and to hold a work platform in a substantially vertical axis. **3637, 3642**

D. Industrial trucks, such as rough terrain forklifts, may be used to elevate and position workers under specific conditions. **3657**

E. Elevators (construction) are required as follows:

1. For structures or buildings 60 ft. or more above ground level or 48 ft. below ground level. **1630(a)**

2. At demolition sites of seven or more stories or 72 ft. or more in height. **1735(r)**

   *Note: Elevators must be inspected and tested in the presence of a DOSH representative before use.*
   *A permit from DOSH to operate is required.* **1604.29(a)**

F. Personnel hoists may be used at special construction sites, such as bridges and dams, if approved by a registered engineer. **1604.1(c)**

G. Ladders can be used to gain access to working surfaces above and below ground level under certain conditions. **1675**

H. Ramps and runways provide means of access for foot or vehicle traffic. **1623, 1624, 1625**

I. Stairways must be installed in buildings that have two or more stories or are 24 ft. or more in height. **1629(a)(1)**

1. For buildings of two and three stories, at least one stairway is required. **1629(a)(4)**

2. For buildings of more than three stories, two or more stairways are required. **1629(a)(4)**

J. The following routes of access are prohibited:

1. Endless-belt-type manlifts. **1604.1(a)(3)**

2. Single-cleat more than 30ft or double-cleat ladders more than 24 ft. long. **1629(c)**

3. Cleats nailed to studs. **1629(b)**

4. Rides on loads, hooks, slings, or concrete buckets of derricks, hoists, or cranes. **1718(a), 1720(c)(3)**

## Administrative Requirements

Employers must meet certain administrative requirements that may include Cal/OSHA notification, specific registration, permitting, certification, recordkeeping, and the posting of information in the workplace. Some of these requirements depend on the construction trade or type of activity in which employers are involved. The more common requirements are listed below:

A. Documents required at the job site include the following:

1. IIP Program: program document may be kept in the office. **1509(a), 3203(a)**

2. Code of Safe Practices. **1509(b)**

3. All Cal/OSHA-required permits. **341**

4. All Cal/OSHA-required certifications. **Various**

5. Respiratory Protection Program, for all work sites where respirators are mandatory. **5144(c)**

6. Heat Illness Prevention. **3395**

7. Fall protection plan, if required. **1671.1**

B.  Postings required at the job site include the following:

1.  Cal/OSHA poster "Safety and Health Protection on the Job". **340**

2.  Code of Safe Practices. **1509(b),(c)**

3.  Emergency phone numbers. **1512(e)**

4.  Employee access to records notification, to show that employees have the right to gain access to medical and exposure records. **3204(g)**

5.  Operating permit for air tanks. **461(a)**

6.  Operating rules for industrial trucks, and tow tractors (if used), where employees operate forklifts. **3664, 3650(c)**

7.  Authorized access, at controlled access zones (CAZs). **1671.1, 1671.2**

8.  Variance process. **411.3**

9.  Cal/OSHA registration. **341.4, 341.10**

10. Citations. **332.4**

11. Hazard warning signs at the following job sites:

    a) Where asbestos work is being done. **341.10, 1529(k)**

    b) Where lead work is being done. **1532.1(m)**

    c) At confined work spaces. **5156, 5157, 5158**

    d) At controlled access zones. **1671.2**

    e) On cranes, concrete pumps, high-lift trucks, etc., (high-voltage warning signs). **2947**, Group 13

    f) On powder-actuated tools. **1691(n)**

    g) On lasers (laser levels, etc.). **1801(d)**

    h) On air compressors with an automatic-start function. **3320**

C. Recordkeeping requirements are included in T8 CCR for the purpose of establishing a historical record of compliance. These requirements include the following:

1. OSHA Log 300.

2. Lock-out/block-out activity records. **3314**

3. Operation and maintenance activity records. **1509, 3203**

4. Medical surveillance program and records.

5. Training records.

6. Inspection records.

D. Reports and notifications to Cal/OSHA must be made of the following incidents and activities:

1. Serious injury or death. A report must be made immediately by telephone (within 8 hours) to a district office. Employers are allowed 24 hours if they can show that circumstances prevented the report from being made in 8 hours. **342(a)**

   *Note: A serious injury or illness is defined as one that requires inpatient hospitalization for more than 24 hours of care other than medical observation or as one in which an employee suffers a loss of a member of the body or a serious degree of permanent disfigurement. 330(h)*

2. Blasting accidents or unusual occurrences. A report must be forwarded to the district office within 24 hours. **5248(a)**

3. Construction activities annual permit. Employers governed by an annual permit must notify DOSH before starting the work. **341.1(f)**

4. Asbestos-related work. The DOSH district office must be notified 24 hours before starting work that is subject to registration. **341.9(a)**

5. Use of regulated carcinogens. The employer must report operations involving the use of a regulated carcinogen within 15 days. **5203**

6. Construction involving lead-work. Written notification must be made to the DOSH district office 24 hours before starting work. **1532.1(p)**

E. Notifications to employees must be made for the following:

1. The employer shall notify affected employees of the monitoring results of asbestos and methylenedianiline within 5 working days following receipt of monitoring results. **1529(f)(5)(A), 1535(f)(7)(A)**

2. The employer shall notify affected employees of the monitoring results of cotton dust and vinyl chloride within 15 working days following receipt of monitoring results. **5190(d)(4)(A), 5210(d)(6)**

F. Project or Annual permits issued by Cal/OSHA are required for various construction activities. **341**

A Project Permit is required for: **341(d)**

» Use of diesel engines in any mine or tunnel.

» Demolition or dismantling a structure more than 36 feet high. **341(d)(3)**

» Erecting/raising/lowering or dismantling a fixed tower crane.

Annual Permit is needed for employers when the structure is over 36 ft high: **341(d)(4)**

» Erection and placement of structural steel or structural members other than steel.

» Installation of curtain walls/precast panels or fascia.

» Installation of metal or other decking.

» Forming or placement of concrete structures/decks on steel structures.

» Installation of structural framing (including roof framing) or panelized roof systems.

Annual or Project Permit is needed for:

» Construction of trenches or excavations 5 feet or deeper into which a person is required to enter.

» Erection and placement of scaffolding vertical shoring, or falsework more than 36 feet high.

Operating permit is required for:

» Operating specified air compressors. **461**

» Operating tower cranes if the employer is subject to **341, 341.1, 344.70.**

*Note: Most permits can be obtained from aDOSH district office. A safety conference and a review of the employer's safety program will be scheduled before permit issuance.* **341.1(c)**

*Exception: Permit requirements do not apply to certain activities. See* **341(e).**

G. Certification requirements are necessary in the following circumstances:

1. Power operated cranes and derricks exceeding 3 tons rated capacity shall not be used in lifting service until the equipment has been certified by a DOSH licensed certifier. **1610.9**

2. Operators of mobile and tower cranes must have valid certificate. See exceptions. **1618.1**

3. Asbestos consultants and site surveillance technicians must be certified by DOSH. **341.15**

4. Training certification is required for many activities and trades (see specific SOs). Title 8

H. Registration and licensing are required in the following circumstances:

1. Asbestos registration. An employer must register with DOSH when engaged in asbestos-related work on 100 sq. ft. or more of surface area. **341.6**

2. Blaster's License. The blaster must be a licensed blaster or directed by a licensed blaster, and be at least 21 years of age. **5238**

## Aerial Devices and Elevating Work Platform Equipment

A. Aerial devices, such as cherry pickers and boom trucks, may be vehicle-mounted or self-propelled and used to position employees. **3637**

General safety requirements are as follows:  **3648**

1. Only authorized persons may operate aerial devices. **3648(c)**

2. Aerial devices must not rest on any structure. **3648(a)**

3. Controls must be tested before use. **3648(b)**

4. Workers must stand only on the floor of the basket. No planks, ladders, or other means are allowed to gain greater heights. **3648(e)**

5. A fall protection system must be worn and attached to the boom or basket. **3648(o)**

6. Brakes must be set when employees are elevated. **3648(g)**

7. An aerial lift truck must not be moved when an employee is on the elevated boom platform except under conditions listed in **3648(l)**.

> **Toolbox**
> "Aerial Lifts"
> **PUB201**   www.oshatools.com

B. Elevating work platform equipment, such as vertical tower, scissor lift, and mast-climbing work platform may be used to position employees and materials. **3642**

General safety requirements are as follows: **3642**

1. The platform deck shall be equipped with a guardrail or other structure around its upper periphery. Where the guardrail is less than 39 in. high, a personal fall protection system is required. **3642(a)**

2. The platform shall have toe boards at sides and ends. **3642(f)(1)**

3. No employee shall ride, nor tools, materials, or equipment be allowed on a traveling elevated platform. See exceptions. **3646**

4. Units shall not be loaded in excess of the design working load. **3646(f)**

C. The following information must be displayed on the device: **3638(c)**

1. Manufacturer's name, model, and serial number.

2. Rated capacity at the maximum platform height and maximum platform travel height.

3. Operating instructions.

4. Cautions and restrictions.

D. Devices must be designed to applicable American National Standards Institute (ANSI) standards. **3638(b)**

*Note: See clearances for operations near high-voltage conductors in the Electrical section of this guide.*

# Airborne Contaminants and Dust

The employer must control employees' exposure to airborne contaminants and employees' skin contact with those substances identified in Table AC-1 of **5155** and **1528**. Airborne contaminants suspended in the air can exist in different forms including gases, vapors and, particulates (particles of either liquids or solids). Table AC-1 contains the Permissible Exposure Limit (PEL) for these substances. The PEL applies to the sum of the exposures to the substance in the vapor state and from the particulate fraction. **5155**

Some of the substances listed in Table AC-1 also have specific performance standards, noted in the CSOs and the GISOs, for controlling employee exposure. These substances include asbestos **(1529)**; cadmium **(1532)**; lead **(1532.1)**; benzene **(5218)**; methylenedianiline **(1535)**; concrete and masonry materials **(1530.1)**; cotton dust **(5190)**; vinyl chloride **(5210)**; and welding fumes **(1536, 1537)**.

> **Toolbox**
> "Exposure to Respirable Silica"
> **PUB223** www.oshatools.com

Airborne contaminants must be controlled by: **5141**

» Applying engineering controls.

» Removing employees from exposure to the hazard and by limiting the daily exposure of employees to the hazard.

» Providing respiratory protective equipment whenever such engineering controls are not practicable or fail to achieve full compliance.

> **Toolbox**
> "Health Effects of Worker Exposure to Asphalt"
> **PUB225** www.oshatools.com

# Air Compressors

General requirements for air compressors include:

A. Employers must obtain a DOSH permit for the air tanks of air compressors operated at a work site. **461(a)**

   *Exception: No permit is required for tanks with a diameter of less than 6 in., tanks equipped with a safety valve set to open at no more than 15 psi pressure, or tanks having a volume of 11/2 cu. ft. or less with a safety valve set to open at no more than 150 psi.* **461(f)**

B. Warning signs are required for electric air compressors equipped with an automatic-start function. **3320**

C. Safety valves must be popped weekly. **1696(d)**

D. Air tanks must be drained per manufacturer's recommendation. **1696(c)**

E. Fans shall be guarded with a shroud or side screens. **1696(b)**

F. Portable air compressors on wheels must be prevented from rolling. **1696(a)**

# Asbestos

The word asbestos refers to six naturally occurring, fibrous, hydrated mineral silicates that differ in chemical composition. They are actinolite, amosite, anthophyllite, chrysotile, crocidolite, and tremolite. Nonfibrous forms of the last three minerals listed here are regulated by GISO **5208.1**. You may encounter asbestos at a construction site in the following applications and areas:

» Excavations where asbestos-bearing rock outcroppings are at or near the surface.

» Fireproofing for steel-frame high-rise buildings.

» Pipe and boiler insulation.

» Insulators of electrical conductors.

» Plaster, cement, drywall, and taping compounds.

» Floor tile and tile adhesives.

» Acoustical ceilings (tiles and sprayed on).

» Asbestos-cement piping, shingles, and panels.

» Roofing felt and sealing compounds.

**Toolbox**

"Asbestos in Construction"

**PUB202** www.oshatools.com

Because asbestos exposure has been linked to serious illnesses, Fed/OSHA and Cal/OSHA have implemented strict regulations to minimize exposures to work site and "take-home" asbestos. Below find a summary of regulatory requirements:

A. Construction projects are subject to regulation under **1529** if they involve one or more of the following activities, regardless of the percentage of asbestos present:

1. Demolition or salvage of structures where asbestos is present.

2. Removal or encapsulation (including painting) of materials that contain asbestos.

3. Construction, alteration, repair, maintenance, or renovation of structures, substrates, or portions thereof that contain asbestos.

4. Installation of products that contain asbestos.

5. Erection of new and the improvement, alteration, and conversion of existing electric transmission and distribution lines and equipment.

6. Excavation that may involve exposure to naturally occurring asbestos, excluding asbestos mining and milling activities.

7. Routine facility maintenance.

8. Transportation, disposal, storage, and containment of and site housekeeping activities involving asbestos or materials containing asbestos.

9. Asbestos spills and emergency cleanups.

Regulatory requirements for work activities subject to **1529** vary depending on the percent, the amount, or the type of asbestos-containing materials involved. Listed below are selected requirements and the activities to which they apply.

B. Cal/OSHA administrative requirements are as follows:

1. Registration and district notification, if disturbing 100 sq. ft. or more of manufactured construction materials containing more than 1/10 of 1% of asbestos-containing construction material (ACCM). **341.6(a)**

2. Carcinogen notification, with exposures in excess of permissible exposure limits (PELs).

3. The employer shall notify affected employees of the monitoring results of asbestos within 5 working days following receipt of monitoring results. **1529(f)(5)(A)**

4. DOSH certification is required for all person performing duties of an asbestos consultant or technician. **341.15(a)**

   "Asbestos consultant" means any person who contracts to provide professional health and safety services relating to asbestos. **1529(q)(1)**

C. Training is required for all employees engaged in Class I through IV work and all work in which they are likely to be exposed in excess of the PELs. The training must be provided:

1. At the employer's expense.

2. Before or at the time of initial assignment.

3. Annually after initial training.

4. In accordance with **1529(k)(9).**

D. Permissible exposure limits (PELs): The employer must ensure that employee exposures do not exceed: **1529(c)**

1. Eight-hour time-weighted average of 0.1 fibers/cc.

2. Thirty-minute excursion limit of 1 fiber/cc. **1529(c)**

E. Multi-employer work sites are regulated under **1529**:

1. The general contractor on the project must exercise general supervisory authority. **1529(d)**

2. An employer doing work involving asbestos must notify other employers at the site. **1529(d)**

3. All employers on site must ensure that their own employees are not exposed to asbestos fibers because of a breach in containment or control methods used by the creating employer. **1529(d)**

F. Exposure assessments and monitoring are required as follows:

1. Initial exposure assessment must be made by all employers subject to **1529** before or at the onset of the project. **1529(f)(2)**

2. Daily exposure monitoring of employees must be conducted by all employers disturbing materials that contain more than 1% asbestos in Class I and II work. **1529(f)(3)**

3. All employers must monitor daily representative exposure of employees performing Class I and II work.

*Exceptions* **1529(f)(3)***:*

*No monitoring required when:*

» The employer has made a negative exposure assessment for the entire operation.

» Employees are equipped with supplied air respirators (SARs) operated in the pressure demand mode, or other positive pressure mode respirator. However, employees performing class I work using certain control methods shall be monitored daily even if they are equipped with SARs.

4. Periodic exposure monitoring of employees must be conducted when disturbing asbestos-containing materials (ACMs) in operations involving other than Class I and II work during which the PELs might be exceeded. **1529(f)(3)**

G. Respirator protection requirements are specific to asbestos-related activities and exposures as outlined in **1529(h)**:

1. The employer must provide appropriate respirators to employees; however, employers shall not use filtering facepiece respirators for use against asbestos fibers. **1529(h)(3)**

2. The appropriate respirator must be selected from **Table 1 of 5144(d)(3)(A)(1). 1529(h)**

3. The employer must provide HEPA filters for powered and non-powered air-purifying respirators. **1529(h)(3)(B)**

4. A written respiratory protection program must be implemented in accordance with **5144(c). 1529(h)(2)**

H. Methods of compliance and work practices are noted below:

1. The wet method must be used unless the employer can demonstrate that it is not feasible. **1529(g)(1)**

2. Vacuum cleaners with high-efficiency particulate air (HEPA) filters must be used to clean up ACM and presumed asbestos-containing material (PACM). **1529(g)(1)**

3. Prompt cleanup and disposal in labeled leak-tight containers are required except as specified in **1529(g)(8)(B). 1529(g)(1)**

4. Specific work practices for different activities are also outlined in **1529. 1529(g)(4-11)**

5. Stripping of finishes shall be conducted using low abrasion pads at speeds lower than 300 rpm and wet methods. **1529(g)**

I. Prohibited work practices and controls are as follows:

1. Spraying of any substance containing any amount of asbestos (see exception). **1528**

2. High-speed abrasive disc saw cutting of ACM or PACM without appropriate point of cut ventilator or enclosures with HEPA filtered exhaust air. **1529(g)(3)**

3. Using compressed air to remove asbestos or materials containing asbestos. **1529(g)(3)**

4. Dry sweeping, shoveling, or other dry cleaning of dust or ACM or PACM debris. **1529(g)(3)**

5. Rotating employees as a means of reducing exposure to asbestos. **1529(g)(3)**

## Blasting (Abrasives/Sand)

Regulations for blasting with abrasives and sand include the following:

A. Employees must wear supplied-air respirators (covering the head, neck, and shoulders):

1. During abrasive blasting when dust may exceed limits specified in **5155.** **5151(b)(1)(B)**

2. During abrasive blasting with silica sand or where toxic material evolves. **5151(b)(1)(C)**

*Note: A dust filter respirator may be used for 2 hours during abrasive blasting if the concentration of silica dust is less than ten times the limit specified in* **5151(b)(1)(C)**.

B. Hearing protection must be worn as required by **1521**.

C. Body protection must be worn as required by **1522**.

**Toolbox**
"Exposure to Respirable Silica"
**PUB223**   www.oshatools.com

# Blasting (Explosives)

A person must hold a valid California Blaster's License and must be physically present when performing, directing, and supervising blasting operations. No person under the age of 21 years shall be permitted in any explosive magazine or be permitted to use, handle, or transport explosives. **5238(a), 5276(g)**

*Exception: Persons 18 years or older and under the direct personal supervision of a licensed blaster.*

A. Blaster's License requirements are discussed in **344.20**.

B. All blasting accidents affecting worker safety must be reported to DOSH within 24 hours. **5248(a)**

   *Note: Accidents involving a serious injury or illness must be reported to DOSH immediately but not longer than 8 hours.* **342(a)**

C. Explosives must be stored in the proper type of magazine (see **5252** Table EX-1). **5251(a)**

D. Caps and detonators must be stored in separate magazines away from other explosives. **5251(b),(c)**

E. Storage requirements are discussed in **5251, 5252, 5253**.

F. Transportation requirements are discussed in **Subchapter 7 Article 115 (Index). 5270**

G. Safety rules for blasting operations are as follows:

   1. No smoking or open flames are permitted within 50 ft. of explosives handling. **5276(a)**

   2. No source of ignition, except during firing, is permitted in areas containing loaded holes. **5276(a)**

   3. Only non-sparking tools are to be used for opening containers of explosives. **5276(b)**

   4. Explosives must be kept clear of electrical circuits by 25 ft. **5276(d)**

5. Unused explosives must be returned promptly to the magazine. **5276(e)**

6. Blasting mats must be used when flying material could damage property. **5276(f)**

7. A tally sheet that records all movement of explosives must be kept at each magazine. **5251(n)**

8. Holes may be loaded only after all drilling is complete (see exception in **5278(a)**). **5278(a)**

9. No vehicle traffic should pass over loaded holes. **5278(c)**

10. Loaded holes must be attended. **5278(o)**

11. Workers must not try to quench an explosive's fire. **5276(l)**

12. Explosives at a blast site must be attended. **5278(o)**

13. No one but the attendant(s), the loading/detonation crew, inspection personnel, and authorized supervisory personnel shall be allowed within 50 feet of the loaded holes. **5278(o)(3),(w)(3)**

14. Blasts shall not be fired without the licensed blaster-in-charge verifying the conditions listed in **5291(b)**, and without a warning signal/procedure. The signals shall be heard clearly in areas that could possibly be affected by the blast. **5291(b)**

# Carcinogens

Whenever carcinogenic (cancer-causing) chemicals, as specified in the GISOs **Article 110 Regulated Carcinogens** are present in construction materials, the employer must comply with the reporting requirements and safety rules. **5203**

1. For all regulated carcinogens that specify a requirement for the employer to establish a regulated area, use of a regulated carcinogen within such a regulated area shall be reported to Cal/OSHA. For regulated carcinogens that do not have a regulated area requirement, use of the regulated carcinogen shall be reported in certain circumstances. **5203(c)**

2. Initial use/changes in reported information of a regulated carcinogen shall be reported to Cal/OSHA in writing within 15 days. **5203(d)**

3. Employers with temporary worksites need to provide the initial use/changes report for their permanent workplace location. **5203(e)**

4. A copy of the applicable written report of use, temporary worksite notification, and emergency report shall be posted for affected employees to see. **5203(g)**

5. In case of emergency: **5203(f)**

   » A report of the occurrence of an emergency and the facts obtainable at that time shall be made to Cal/OSHA within 24 hours.

   » A written report shall be filed within 15 days.

The material safety data sheet (MSDS) and labels on the container must be reviewed to determine the presence of carcinogens.

## Code of Safe Practices

The Code of Safe Practices is a set of work site rules that stipulate how to perform job duties safely and to keep the work site safe. The following are selected requirements:

A. The employer must develop and adopt a written Code of Safe Practices. **1509(b)**

   *Note: Plate A-3 in Appendix A of **1938** is a suggested code. The code is general and should be used as a starting point for developing a code that fits the contractor's operations.*

B. It must be specific to the employer's operations. **1509(b)**

C. It must be posted at each job site office or be readily available at the job site. **1509(c)**

D. Workers, when first hired, shall be given instructions regarding the hazards and safety precautions and directed to read the Code of Safe Practices. **1510(a)**

E. Supervisors shall conduct "toolbox" or "tailgate" safety meetings, or equivalent, with their crews at least every 10 working days to emphasize safety.

## Competent Person

A competent person is defined in **1504(a)** as one who is capable of identifying existing and predictable hazards in the surroundings or working conditions that are unsanitary or dangerous to employees. The competent person has authority to impose prompt corrective measures to eliminate these hazards.

Some SOs identify specific requirements for the competent person's training, knowledge, abilities, and duties. Following is a list of CSOs that require the use of a competent person: (1) asbestos **1529(o)**; (2) excavation **1541, 1541.1**; (3) cadmium **1532(b)**; (4) fall protection **1670, 1671.2**; (5) bolting and riveting **1716**; (6) pressurized worksites **6075** and (7) lift-slab construction operations **1722.1(i)**.

# Concrete Construction

Injuries and illnesses common to the concrete construction industry are as follows:

» Burns, rashes, and skin irritations from exposure to cement dust or wet concrete.

» Silicosis, a respiratory disease caused by inhaling silica dust, from exposure to concrete dust during such operations as concrete cutting, drilling, grinding, or sandblasting.

» Broken bones, lacerations, and crushing injuries caused by falls from elevated work surfaces; impalement by rebar or other objects; and impact from falling objects, form and shoring failure, and structural failure of components of the project.

Because the hazards associated with concrete construction are great, employees must use appropriate personal protective equipment and conform to safe work practices at all times (see below).

A. Placement of Concrete. **1720**

1. Concrete pumping equipment and placing booms shall be set-up and operated according to manufacturer's guidelines and the Title 8 Safety Orders.

2. The manufacturer's operation manual shall be maintained in legible condition and shall be available at the job site.

3. Controls in the equipment shall have their function clearly marked.

4. Operation of concrete placing booms in proximity of overhead high-voltage lines shall be in accordance with Article 37 of the High-Voltage Electrical Safety Orders.

5. Equipment shall be inspected by a qualified operator prior to daily use and the inspection must be documented.

B. Forms/falsework and vertical shoring in the Forms, Falsework, and Vertical Shoring section of this guide. **1717**

C. Masonry construction. **1722**

1. All masonry walls more than 8 ft. high must be braced to prevent overturning and collapse unless the wall is adequately supported through its design or construction method. The bracing shall remain in place until permanent supporting elements of the structure are in place. **1722(b)**

2. A limited access zone (LAZ) shall be established whenever a masonry wall is being constructed and must conform to the following:

   a) The LAZ shall be established before the start of construction. **1722(a)(1)**

   b) The LAZ shall be established on the unscaffolded side. **1722(a)(2)**

   c) The width of the LAZ shall be equal to the height of the wall to be constructed plus 4 ft. and shall run the entire length of the wall. **1722(a)(3)**

   d) The LAZ shall be entered only by employees actively engaged in constructing the wall. No other employee shall be permitted entry. **1722(a)(4)**

   e) The LAZ shall remain in place until the wall is adequately supported to prevent collapse unless the height of the wall is more than 8 ft., in which case the LAZ shall remain in place until the requirements of **1722(b)** have been met. **1722(a)(5)**

D. Precast, prefabricated concrete construction, tiltup, panels. **1715**

1. An erection plan, addenda, and procedure shall be prepared by or under the direction of a Professional Engineer registered in California.

2. The erection plan, addenda, and procedure shall be available at the job site.

3. Job site inspections shall be made by the responsible engineer (or representative) during the course of erection.

4. Proposed field modifications shall be approved by the responsible engineer.

**Toolbox**

"Concrete Construction"

**PUB203**   www.oshatools.com

E. Rebar and other impalement hazards. **1712**

1. Employees working at grade or at the same surface level as exposed protruding rebar or similar projections shall be protected against impalement by guarding exposed ends that extend up to 6 feet above grade or other work surface, with approved protective covers or troughs (see illustrations 1 and 2). **1712(c)**

2. Employees working above grade or above any surface and who are exposed to protruding rebar or similar projections shall be protected from impalement by:

   a) The use of guardrails, or

   b) Approved fall protection systems, or

   c) Approved troughs and covers per **344.90**, **1712(c)**.

3. Job-built wood protective covers and troughs shall be built of at least "standard- grade" Douglas Fir.

4. Manufactured protective covers shall be approved by Cal/OSHA, per **344.90.**

5. Personal fall protection must be used while employees place or tie rebar in walls, columns, piers, and other structures more than 6 ft. high. **1712(e)**

   *Exception: Personal fall protection is not required during point-to-point horizontal or vertical travel on rebar up to 24 feet above the surface below if there are no impalement hazards.*

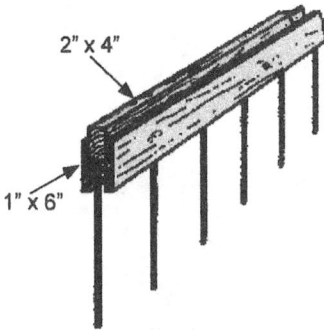

**Illustration 1**
**Troughs**

2" x 4"

1" x 6"

Protuding
reinforced steel

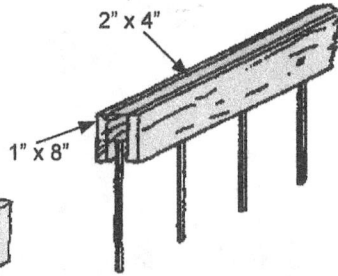

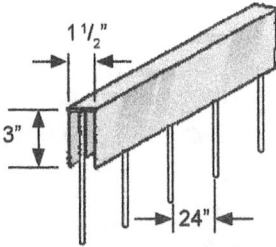

2" x 4"

1" x 8"

1 $1/_2$"

3"

24"

**14 Guage Steel Trough**

Troughs can be used for impalement protection providing the following applies:

» The trough designs shown above can be used when employees are working at heights of 6 ft. or less "above grade".

» If employees are working at heights above 6 ft., the design must be specified by an engineer (Ca PE).

» Job-built wood troughs must be constructed of at least "standard grade" Douglas Fir.

**Toolbox**
"Ergonomic Survival Guide–Cement Masons"
**PUB252**   www.oshatools.com

# Illustration 2
## Protective Covers

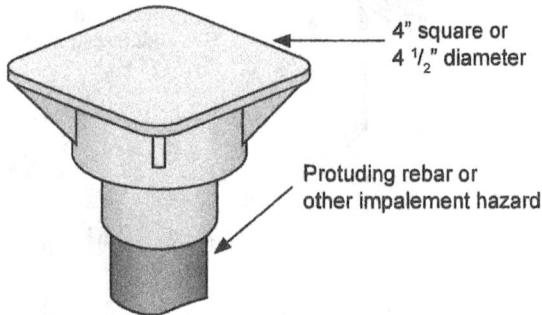

4" square or
4 $\frac{1}{2}$" diameter

Protuding rebar or
other impalement hazard

Manufactured protective covers used for impalement protection must meet the following requirements:

» The protective covers must be Cal/OSHA approved.

» The cover surface must be at least 4 in. square. If the cover is round, its surface must have a minimum diameter of 4 $\frac{1}{2}$ in. For a trough, the protective cover must be at least 4 in. wide.

» The protective covers used "above grade" must be designed to withstand the impact of a 250 lbs. weight dropped from 10 ft.

» The protective covers used "at grade" must be designed to withstand the impact of a 250 lbs. weight dropped from 7 $\frac{1}{2}$ ft.

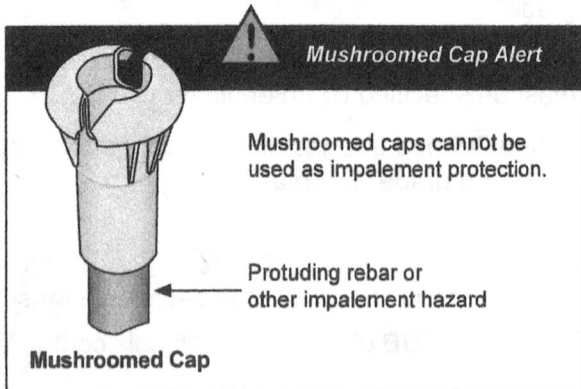

### Mushroomed Cap Alert

Mushroomed caps cannot be used as impalement protection.

Protuding rebar or
other impalement hazard

**Mushroomed Cap**

6. Guying and supporting of all rebar for walls, piers, columns, and similar vertical structures are required.

7. Wire mesh rolls shall be secured to prevent dangerous recoiling action. **1712**

F. Concrete finishing

1. Powered finishing tools must be equipped with a deadman-type control.

2. Bull float handles must be constructed of a nonconductive material if they could come into contact with energized electrical conductors.

## Confined Spaces

Every year several confined space entrants and would be rescuers die from hazards, such as oxygen deficiency, toxic and explosive atmospheres, and uncontrolled energized equipment. To prevent such accidents employers must be able to:

» Recognize a confined space and the specific hazards associated with that space.

» Know and understand T8 CCR **5156, 5157, 5158** and related requirements concerning respiratory protection, fall protection, lock-out/block-out procedures, fire prevention, and rescue.

» Implement the safety orders effectively.

*Note: For most construction work* **5158** *applies; however, work in confined spaces during refurbishing operations may be subject to the permit-required confined space regulations in* **5157** *(see* **5156**).

A. Confined space is defined in **5158(b)(1)** as space that exhibits both of the following conditions:

1. The existing ventilation does not remove dangerous air contaminants or oxygen-deficient air that exists or may exist or develop.

2. Ready access or egress for the removal of a suddenly disabled employee is difficult because of the location or size of the opening(s).

> **Toolbox**
>
> "Is It Safe to Enter a Confined Space?"
>
> **PUB259** www.oshatools.com

B. The following are examples of some of the locations which may exhibit confined-space conditions:

1. Trenches and excavations

2. Sewers and drains

3. Tanks

4. Vaults

5. Wells and shafts

6. Crawl spaces

7. Ducts

8. Compartments

9. Pits, tubs, and bins

10. Pipelines

C. Employers must check initially and if conditions can change, employers must check on an ongoing basis to discern whether work locations exhibit confined-space conditions.

» If confined-space conditions have been identified, the following must be completed before employees may begin work:

1. Written operating procedures must be prepared, and employees must be trained. **5158(c)(1),(2)**

2. Lines containing hazardous substances must be disconnected, blinded, or blocked. **5158(d)(1)**

3. The space must be emptied, flushed, or purged. **5158(d)(2)**

---

4. The air must be tested for dangerous contamination or oxygen deficiency. **5158(d)(5)(A)**

5. Ventilation must be provided if testing reveals any atmospheric hazard. **5158(d)(6)**

D. Working in a confined space where dangerous air contamination exists requires:

1. Appropriate respiratory protection. **5158(d)(11), 5158(e)(1)**

2. Provisions for ready entry and exit where feasible. **5158(d)(10)**

3. The wearing of a safety harness attached to a retrieval line and retrieval equipment (Illustration 3). **5158(e)(1)(C),(E)**

   *Exception: See* **5158(e)(1)(C)**

**Illustration 3**

**Retrieval Equipment in Use**

4. One standby employee (with entry gear) trained in first aid and cardiopulmonary resuscitation plus one additional employee within sight or call. **5158(e)(1),(2)**

5. Effective means of communication between the employee in the confined space and the standby employee. **5158(e)(2)**

6. Ongoing atmospheric testing for dangerous air contamination and oxygen deficiency. **5158(d)**

7. Ongoing surveillance of the surrounding area to avoid hazards, such as vapors drifting from nearby tanks, piping, sewers, and operations. **5158(c)(1)(B)**

## Corrosive Liquids

Employers must provide the following when employees handle corrosives:

» Personal protective equipment. **1514(a)**

» Material safety data sheet (MSDS) in English. **5194(g)**

» Properly labeled containers with appropriate hazard warnings. **5194(f)**

*Note: Employers who become newly aware of any significant information regarding the hazards of a substance shall revise the labels for the substance within three months of becoming aware of the new information.*

» An eyewash and a deluge shower that meet ANSI standards. **3400(d), 5162**

*Note: Emergency eyewash facilities and deluge showers shall be in accessible locations that require no more than 10 seconds for the injured person to reach.*

» A written hazard communication (haz-com) program. **5194(e)**

# Cranes

Hazards associated with crane operations are electrocution from overhead power lines and equipment failures because of operator error; faulty or damaged equipment; overloading; support failure such as ground or outrigger collapse; and miscommunication.

All of the regulations for cranes used in construction became effective July 7,2011 and are covered in T8 CCR Sections **1610-1619, 1694, 2940,** and **6060**. **1610-1619** covers Cranes and Derricks in Construction, **1694** covers Side Boom Cranes, **2940** covers Mechanical Equipment, and **6060** covers Procedures During Dive.

Employers and employees, in order to maintain safe and healthful working conditions, must ensure that:

1. Manufacturer's instructions are followed.

2. All crane operators have a valid certificate of competency for the specific type of crane that they are operating.

3. Necessary tools, protective equipment, and trainings are provided.

4. Employees comply with all requirements of crane operation and perform tasks safely at all times.

Below is a summary of the regulatory requirements for cranes and derricks used in construction:

A. General requirements for cranes and derricks are given in the Subsections within Section **1610**. Requirements include:

1. Scope - applies to power operated equipment, when used in construction that can hoist, lower and horizontally move a suspended load. **1610.1.** This article does not apply to: power shovels, excavators, wheel loaders and backhoes; even when used with chains, slings or other rigging to lift suspended loads. See **1610.1(c)** for full list of excluded equipment.

2. Design requirements are given in **1610.2** and **4884**.

3. Definitions as per **1610.3**.

4. Design, construction and testing of cranes and derricks, relative to equipment manufacture date, pre/post July 7, 2011, with over 2000 lbs of hoisting/lifting capacity must meet requirements in **1610.4**.

5. Ground conditions including slope, compaction, and firmness, and all supporting materials such as blocking, mats, cribbing, marsh buggies etc. must meet the requirements in **1610.5**, to include manufacturer's specifications for adequate support and degree of level per **1610.5(b)**.

6. Equipment modifications or additions which affect the capacity or safe operation of the equipment are prohibited except where the requirements of subsections as shown in **1610.6** are met.

7. Fall protection is critical in crane operations and must be provided by employers. The fall protection system varies depending on the type of crane being used and the work activity. Requirements for fall protection are given in **1610.7**.

8. For cranes with a rated hoisting/lifting capacity of 2,000 pounds or less, the employer must ensure that all of the requirements in **1610.8** are met. If manufacturer's operating instructions are unavailable, the employer shall develop procedures necessary for the safe operation of the equipment and attachments, per **1610.8(c)(2)(A)**.

9. For cranes with a rated hoisting/lifting capacity over 3,000 pounds, the employer must ensure that the cranes, derricks and accessory gears are not used until there is a verification of current annual certification as per **1610.9**.

B. Section **1611** and its subsections **1611.1** through **1611.5** address all of the safety requirements related to assembly and disassembly operations.

1. When assembling or disassembling equipment (or attachments), the employer must comply with all applicable manufacturer prohibitions and requirements in **1611.1.**

2. The general requirements for assembly and disassembly operations including supervision, competent and qualified person requirements, review of procedures, crew instructions, etc. are given in **1611.2.**

3. Employers/operators must also follow the requirements for dismantling of booms and jibs as specified in **1611.3.**

4. Employer procedures for assembly/ disassembly shall be developed by a qualified person. **1611.4**

5. The employer shall follow the power line safety (up to 350 kV) requirements of **1611.5.** **Employers and employees always need to presume that power lines are energized.**

C. Power line safety is regulated under T8CCR **1612** and its subsections. The requirements vary depending on the voltage of the power line. The following requirements apply:

1. For equipment operations with potential involvement of power lines up to 350 kV, employer shall follow the power line safety requirements, defined work zone and special requirements if closer than 20 ft. to power line, per **1612.1.**

2. For power lines over 350 kV, the employer shall follow all of the requirements, defined work zone and special requirements if closer than 50 ft. to power line, per **1611.5** and **1612.1.** See exceptions.

3. For all energized power lines (all voltages), whenever equipment operations including load lines or loads are closer than the minimum approach distance under Table A, the employer shall prohibit these operations. **1612.3**

4. While traveling under or near power lines with no load, employer must establish procedures and criteria, and follow the safety requirements of T8CCR **1612.4.**

D. Requirements for inspections of cranes and derricks are given in T8CCR **1613**. Specific requirements include:

1. Prior to initial use, all equipment that has modifications or additions which affect the safe operation of the equipment or capacity, shall be inspected by a certificating agency. The inspection shall meet the requirements of T8CCR **1613.1**.

2. Inspections of repaired/adjusted equipment are subject to the requirements, must be qualified person, per **1613.2**.

3. Post assembly inspections are subject to the requirements in **1613.3**.

4. The inspections done each shift are subject to the requirements in **1613.4**.

5. Periodic inspections shall be conducted at least four times a year. Cranes shall not be operated more than 750 hours, between periodic inspections. The inspection shall include all items as per **1613.5**.

6. Annual/Comprehensive inspections need to be done as per **1613.6**.

7. Where there is a reasonable probability of damage or excessive wear, the employer shall stop using the equipment and a qualified person shall inspect the equipment for structural damage, and the causing items/ conditions. **1613.7**

8. Equipment that has been idle for 3 months or more shall be inspected by a certificating agency or qualified person as per T8CCR **1613.5**, before initial use. **1613.8**

9. General inspections must comply with **1613.9**.

10. Inspections of wire ropes are subject to the requirements of **1613.10**.

E. Requirements for the selection and installation of wire ropes are given in **1614**. Selection and installation of original and replacement wire rope shall be as per the wire rope manufacturer, the equipment manufacturer, or a qualified person.

F. Requirements for the safety devices and operational aids are given in **1615** and include:

1. Safety devices such as crane level indicator, horn, jib stops, boom stops etc. are required on all equipment unless otherwise specified. **1615.1**

2. Operational aids such as boom hoist limiting device, boom angle, boom length indicator, load weighing device, etc. are required on all equipment unless otherwise specified. **1615.2**

   *Note: Operational aids are classified into Category I and Category II.* **1615.2**

G. Requirements for the operation of cranes and derricks are given in T8CCR **1616** and include:

1. The employer shall follow manufacturer procedures for operation of the equipment including the use of attachments. Where procedures for operation are unavailable, the employer shall comply with **1616.1**.

   *Note: While operating equipment, devices such as cell phones shall not be used for any other activities (texting, talking etc.) other than signaling.*

2. Whenever there is a concern as to safety, the operator shall have the authority to stop and refuse to handle loads until a qualified person has determined that safety has been assured. **1616.2**

3. Work area control including protecting employees in hazardous areas, communication among operators and signal persons shall be followed as per **1616.3**.

4. Operations shall be conducted and the job controlled in a manner that will avoid exposure of employees to the hazard of overhead loads. Wherever loads must be passed directly over workers, occupied work spaces or occupied passageways, safety type hooks or equivalent means of preventing the loads from becoming disengaged shall be used. All requirements under **1616.4** shall also be met.

5. Boom free fall is prohibited in each of the circumstances mentioned in **1616.5**. Controlled load lowering is required and free fall of the load line hoist is prohibited in each of the circumstances mentioned in **1616.5(d)**.

6. The use of equipment to hoist employees is prohibited except where the employer demonstrates that the erection, use, and dismantling of conventional means of reaching the work area, would be more hazardous, or is not possible because of the project's structural design or worksite conditions. **1616.6(a)**

7. Hoisting of personnel using cranes is possible only when all of the requirements of **1616.6** are met.

   *Note: The requirements of **1616.6** are supplemental, and apply when one or more employees are hoisted.*

8. Supplemental requirements for Multiple-Crane/ Derrick Lifts are provided in **1616.7**. Before beginning a crane/ derrick operation in which multiple crane/derrick will be supporting the load, the operation shall be planned as per **1616.7(a)** and directed by a qualified person.

H. The general requirements for using signals during the operation of cranes and derricks are given in **1617** and include:

1. A signal person shall be provided in each of the situations given under **1617.1**. Only qualified persons shall be permitted to give signals except for a stop signal. Signals to operators shall be by hand, voice, or audible and as per **1617.1**. Recommended hand signals are shown in Illustration 4 on next page.

2. The devices transmitting signals shall be tested on site before start of operations and the devices/ signaling shall meet requirements in **1617.2**

3. Follow the additional requirements in **1617.3** for voice signals.

*Note: Employees shall not text or talk unless it is for signaling purposes.*

# Illustration 4

## Recommended Hand Signals

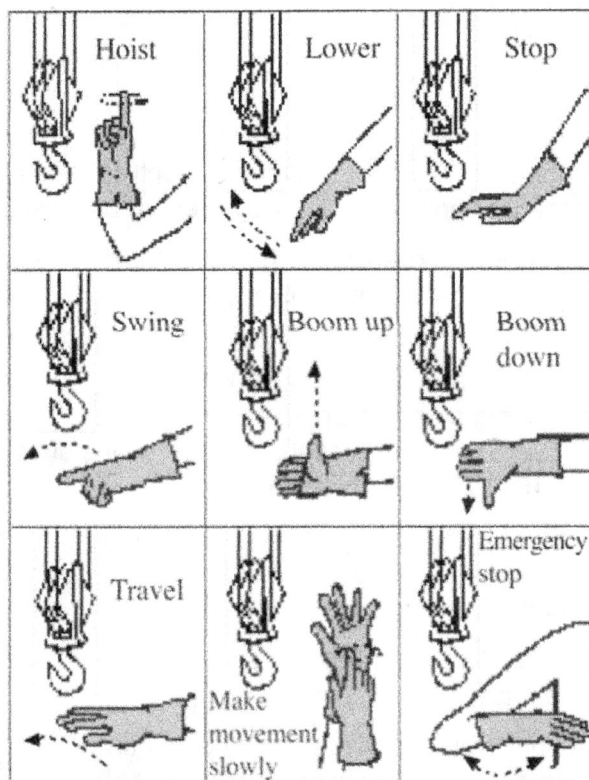

I. The requirements for operator qualification, training and certification are given in **1618** and its subsections. They include:

  1. Operator qualifications/certification/in-training must comply with **1618.1**.

  2. Make sure that each signal person meets the qualification requirements in **1618.2** prior to giving any signals.

  3. Maintenance, inspection and repair personnel are permitted to operate the equipment only where all of the requirements of **1618.3** are met.

4. The employer shall provide training to all operators, signal persons, spotters, competent/qualified persons, and operators-in-training on their specific jobs as per **1618.4**.

J. T8 CCR **1619** has supplemental requirements for certain types of cranes and derricks. Supplemental requirements include:

1. Section **1619.1** contains supplemental requirements for erecting, climbing, operating, dismantling, and all other operations and devices used in regard to tower cranes.

2. The supplemental requirements for derricks, whether temporarily or permanently mounted, are given in **1619.2**.

3. Section **1619.3** contains supplemental requirements for floating cranes/derricks and land cranes/derricks on barges, pontoons, vessels or other means of flotation. See **1619.3** for complete requirements.

4. Overhead and gantry cranes, whether permanently or temporarily installed, are subject to the requirements of **1619.4**.

5. The supplemental requirements for dedicated pile drivers are given in **1619.5**.

K. Side-boom cranes mounted on wheel or crawler tractors shall meet all of the requirements of **1694(d)**.

L. A crane/derrick, used to get divers in/out of water, shall not be used for other purpose until all divers are back on board. **6060**

---

**Toolbox**
"Cranes and Derricks in Construction"
**PUB267**   www.oshatools.com

---

## Demolition

The primary hazards associated with demolition are (1) falls from elevated work surfaces; (2) exposure to hazardous air contaminants; (3) being struck by falling or collapsing structures; and (4) electrical hazards. Regulations to address these hazards include the following:

A. A DOSH permit is required for demolition of any building or structure more than 36 ft. high. The Project Administrator shall hold a Project Permit and all other employers directly engaged in demolition or dismantling activity shall hold an Annual Permit. **341(d)(3)**

B. A predemolition survey must be made to determine whether the planned work will cause:

1. Any structure to collapse. **1734(b)(1)**

2. Worker exposure to hazardous chemicals, gases, explosives, flammable materials, or similarly dangerous substances.**1735(b)**

3. Worker exposure to asbestos. **1529(k)(1), 1735(b)**

4. Worker exposure to lead. **1532.1(d)(1)**

5. Worker exposure to carcinogenic (cancer-causing) chemicals, as specified in GISOs Article 110 Regulated Carcinogens. **5203**

6. Worker exposure to silica. **5144**

C. Utilities to the structure being demolished must be turned off or protected from damage. **1735(a)**

D. Demolition techniques include the following:

1. Entrances to multi-story buildings must be protected by a sidewalk shed or a canopy. **1735(j)**

2. The demolition work on floors and exterior walls must progress from top to bottom. **1735(f)(1)**

   *Exception: Demolition with explosives and for cutting chute holes is not required to progress from top to bottom. 1735(f)(1)*

3. The employer must check continually for hazards created by weakening of the structure's members. If such hazard occurs, it must be corrected before workers may continue. **1735(d)(4)**

4. Floor openings must have curbs and stop logs to prevent equipment from running over the edge. **1735(v)**

5. Wall openings must be guarded except on the ground floor and the floor being demolished. **1735(k)**

6. Walkways not less than 20 in. wide must be provided as a means of access across joists, beams, or girders. **1735(h)**

7. Demolition debris must be kept wet to prevent dust from rising or other equivalent steps taken. **1735(t)**

8. Whenever waste material is dropped to any point lying outside the exterior walls of the building, enclosed chutes shall be used unless the area is effectively protected by barricades, fences, or equivalent means. Signs shall be posted to warn employees of the hazards of falling debris. **1736(a)**

9. Chutes or chute sections that are at an angle of more than 45° from the horizontal must be entirely enclosed except for openings equipped with closures at or about floor level for the insertion of materials. **1736(f)**

10. When chutes are used to load trucks, they must be fully enclosed. Gates must be installed in each chute at or near the discharge end. A qualified person must be assigned to control the operation of the gate and the backing and loading of trucks. **1736(b)**

11. Any chute opening into which employees dump debris by hand must be protected by a guardrail. **1736(d)**

12. When debris is dropped through holes in a floor without the use of chutes, the area onto which the material is dropped shall be completely enclosed with barricades not less than 42 in. high and not less than 6 ft. back from the projected edge of the opening above. Signs that warn of the hazard of falling materials shall be posted at each level. Removal of debris shall not be permitted in the lower drop area until handling of debris ceases above. **1736(f)**

E. Crane demolition work is guided by these regulations: **4941**

1. The wrecking ball's weight must not exceed 50% of the clamshell rating or 25% of the rope-breaking strength. **4941(a)**

2. The swing of the boom should be limited to 30° left or right. **4941(b)**

3. The wrecking ball must be attached with a swivel-type connection. **4941(b)**

4. The load line and ball must be inspected at least twice each shift. **4941(c)**

5. Outriggers are required when using a wrecking ball (truck cranes). **4941(d)**

    *Note: See crane standards in the Cranes section of this guide.* **Group 13 in the GISOs**

## Dust, Fumes, Mists, Vapors, and Gases

Oxygen deficient atmospheres or harmful dusts, fumes, mists, vapors, or gases in concentrations sufficient to present a hazard to employees must be controlled when possible by removing the employee from the exposure, limiting daily exposure, or applying engineering controls. **1528**

A. Whenever the above controls are not practical or fail to achieve full compliance, respiratory protection must be used, according to **5144**. **1528(a)**

B. Ventilation must comply with Article 4 in the GISOs if it is used as an engineering control method. **1528(c)**

C. Common sources of the above hazards may include:

1. Engine exhaust emission (carbon monoxide, NOx, polycyclic aromatic hydrocarbons, and others).

2. Blasting ($CO_2$, NOx, asbestos, silica, dust).

3. Concrete and rock cutting (asbestos, silica, dust).

4. Fuel storage tanks (harmful vapors).

5. Lead abatement (lead particles, lead compounds).

6. Asbestos abatement (asbestos fibers).

7. Demolition (asbestos, silica, lead, dust, etc.).

8. Welding (fumes).

9. Painting and spraying (solvent, vapors, lead).

10. Sand blasting (asbestos, silica, lead, dust).

11. Harmful dust, fumes, mists, vapors, and gases from other sources.

## Electrical

Each year a large number of employees are injured or killed because they come into contact with energized electrical wiring or equipment. The Electrical Safety Orders (ESOs) are designed to control or to eliminate these often deadly exposures and include:

A. General requirements for protection from electric shock (other than excavations). **1518**

   1. The employer must:

      a) Identify exposed or concealed energized electric power circuits if any person, machine, or tool might come into contact with the circuit.

      b) Advise employees of the location of energized circuits, the hazards, and protective measures.

      c) Provide legible markings or warning signs to indicate the presence of energized electrical circuits.

   2. Protective equipment or devices must be used to protect employees if a recognized hazard exists.

   3. When protective insulating equipment is used, it shall comply with the Electrical Safety Orders.

4. Barricades shall be used in lieu of other protective equipment.

*Note:* **1518(d)** *applies to electrical installations present on the jobsite which do not involve excavations. For electrical installations involving excavations as defined in* **1540**.

B. General requirements for low-voltage systems (<= 600 V)

1. Only qualified persons may work on electrical equipment or systems. **2320.1(a)**

2. Maintenance of electrical installations is required to ensure their safe condition. **2340.1**

3. Electrical equipment and wiring must be protected from mechanical damage and environmental deterioration. **2340.26, 2340.11(a)(2), 2340.23**

4. Covers or barriers must be installed on boxes, fittings, and enclosures to prevent accidental contact with live parts. **2340.17(a)**

C. Main service equipment

Whenever the electric utility provides service via overhead lines, the installation must:

1. Consist of an acceptable service pole. **2405.3**

2. Be suitably grounded. **2395.5(b)**

3. Provide suitable overcurrent protection. **2390.1**

D. Wiring methods and devices

1. Flexible cords may be used in place of permanent wiring methods for temporary work if the cords are equipped with an attachment plug and energized from an approved receptacle. **2500.7(a),(b)**

2. Flexible cords must be Type S and cannot be spliced unless they are size No. 12 (or larger). **2500.9(a)**

3. Skirted attachment plugs must be used on all equipment operating at more than 300 V. **2510.7(b)**

*Exception: Plugs or connectors so designed that the arc will be confined within the body or case of the device shall be acceptable.*

E.  Grounding

   1.  Each receptacle must have a grounding contact that is connected to an equipment grounding conductor. **2510.7(a)**

   2.  Temporary wiring must be grounded. **2405.2(g)**

   3.  Electrically powered tools and electrical equipment with exposed, non current-carrying metal parts must be grounded. **2395.45(b)**

      *Exception: Double insulated powered tools need not be grounded.* **2395.45(b)**

   4.  The frame of a portable generator and the frame of a vehicle where the generator is located need not be grounded under certain conditions. **2395.6**

   5.  A system conductor shall be bonded to the generator frame where the generator is a component of a separately derived system. **2395.6(c)**

   **Toolbox**
   "Ergonomic Survival Guide for Electricians"
   **PUB253**   www.oshatools.com

F.  Ground-fault circuit interrupters (GFCIs)

The GFCI device senses ground faults (accidental electrical paths to ground) in circuits and immediately cuts off all electrical power in that circuit.

   1.  GFCIs are required on receptacles that are not connected to the site's permanent wiring and that have a rating of 15 or 20 amps, 120 V, AC, single phase. **2405.4(c)**

   **Toolbox**
   "Electrical Safety"
   **PUB221**   www.oshatools.com

2. The assured equipment grounding conductor program (AEGC program) is an approved alternative to the GFCI requirement if the following program elements are included: **2405.4(d)**

   a) A description of the program must be written.

   b) The employer shall designate one or more qualified persons to implement the program.

   c) Daily visual inspection of included equipment must be conducted.

   d) The following tests shall be performed:

      (1) All equipment grounding conductors shall be tested for continuity and shall be electrically continuous.

      (2) All plugs and receptacles must be tested for proper attachment to the equipment grounding conductor.

**Toolbox**

"Controlling Electrical Hazards"

**PUB206**   www.oshatools.com

   e) The tests shall be performed as follows:

      (1) Before the first use of newly acquired equipment.

      (2) Before equipment is returned to service.

      (3) Before equipment is used after an incident that may have caused damage.

      (4) At intervals not to exceed three months.

   f) The employer shall not make available or permit the use of equipment that has not met the requirements of **2405.4(d)**.

   g) A means of identifying tested equipment shall be provided.

G. High-voltage power lines (> 600 V)

1. Great care must be taken when working or operating equipment near overhead high-voltage power lines.

2. The required minimum safe distances (clearance) from overhead lines energized by 600 V to 50,000 V are: **2946**

   a) For boom-type equipment in transit, 6 ft.

   b) For boom-type equipment in operation, 10 ft.

   c) For people working near overhead lines, 6 ft.

   *Note: See **2946** for minimum required clearances from voltages greater than 50,000 V.*

3. Amusement rides or attractions shall not be located under or within 15 ft. (4.57 m) horizontally of conductors operating in excess of 600 volts. **2946(b)(2)**

4. The following activities are prohibited unless overhead power lines have been de-energized and visibly grounded:

   a) Work over high-voltage lines. **2946(b)(1)**

   b) Work within required clearances. **2946(b)(2)**

   *Note: When work is to be performed within minimum required clearances, the operator of the high-voltage line must be notified by person or persons responsible for the work before proceeding with any work which would impair the aforesaid clearance. **2948***

H. High-voltage warning signs. **2947**

High-voltage warning signs must be posted in plain view of equipment operators.

I. Lock-out procedures

Lock-out procedures must be followed during the cleaning, servicing, or adjusting of machinery. GISO **3314**, ESO **2320.4**

## Elevators, Lifts, and Hoists

Construction elevator and personnel hoist requirements are as follows:

A. An elevator is required for structures or buildings 60 ft. or more above ground level or 48 ft. below ground level. **1630(a)(1)**

B. An elevator is required at demolition sites of seven or more stories or 72 ft. or more in height. **1735(r)**

C. Use of endless-belt-type manlifts is prohibited. **1604.1(a)**

D. Before use, construction elevators must be inspected and tested in the presence of a DOSH representative. A permit to operate is required. **1604.29(a)**

E. Ropes must be inspected at least once every 30 days, and records of these inspections must be kept. **1604.25(j)**

F. A capacity plate must be posted inside the car. **1604.21(b)**

G. Elevators must be operated only by competent, authorized persons. **1604.26(c)**

H. Installation must comply with **1604**.

I. Landings must be provided at the top floor and at least at every third floor. **1630(d)**

J. Landing doors must be mechanically locked so that they cannot be opened from the landing side. A hook-and-eye lock is prohibited. **1604.6(b)**

K. For hoists located outside of a structure, the hoistway enclosures must be 8 ft. high on the building side or the scaffold side at each floor landing and 8 ft. high on all sides of the pit. **1604.5(c)**

L. Hoistway doors shall be at least 6 1/2 ft. high. Solid doors must contain a vision panel. (See **1604.6(a)** for specific requirements.). **1604.6(a)**

M. During inspection and maintenance, the car shall be operated in the slowest speed. In-car operating devices shall not function when car top operation is selected. The car top operating devices shall include an emergency stop button. The tops of cars shall be enclosed by a standard guardrail and toeboard per **3209. 1604.24**

*Exception: See **1604.24(a)(3)(D)***

## Emergency Medical Services

Emergency Medical Services (EMS) must be readily available. **1512(a), (e)**

A. A first aid kit must be provided by each employer on all job sites and must contain the minimum of supplies as determined by an authorized licensed physician or as listed in **1512(c)**. The contents of the first-aid kit shall be inspected regularly to ensure that the expended items are promptly replaced. **1512(c)(1)**

B. Trained personnel in possession of a current Red Cross First Aid certificate or its equivalent must be immediately available at the job site to provide first aid treatment. **1504(a)**, **1512(b)**

C. Each employer must ensure that its employees have access to emergency medical services at the job site. Where more than one employer is involved in a single construction project on a given construction site, the employers may agree to ensure employee access to emergency medical services for the combined work force present at the job site. **1512(a)**

D. Each employer shall inform all of its employees of the procedure to follow in case of injury or illness. **1512(d)**

E. The employer shall have a written plan to provide emergency medical services. **1512(i)**

F.  Proper equipment for prompt transport of the injured or ill person to an EMS facility or an effective communication system for calling an emergency medical facility, ambulance, or fire service must be provided. Telephone numbers for listed emergency services must be posted (see Cal/OSHA poster S-500). **1512(e)**

G.  The employers on the project may agree to ensure employee access to emergency medical services for the combined work force present at the job site. **1512(a)**

H.  Exposure to blood borne pathogens is considered a job-related hazard for construction workers who are assigned first aid duties in addition to construction work. Although construction employers are specifically exempted from GISO **5193** requirements, they are required to provide appropriate protection for employees who may be exposed to blood borne pathogens when providing first aid. **3203**

## Engine Exhaust Emission

Extreme care must be taken when engine exhaust can build up in work spaces, such as confined spaces, excavations, and trenches.

A.  Exhaust purifier devices approved by DOSH or the California Air Resources Board (CARB) must be used to maintain concentrations of dangerous gases or fumes below maximum acceptable concentrations if natural or forced dilution ventilation and exhaust collection systems are inadequate. **5146**

    *Note: Approval by DOSH will be based on the Maximum Allowable Standards for Internal Combustion Engine Exhaust Emissions as set forth in **5146(c)**.*

B. Use of internal combustion engines in tunnels is prohibited. **7070(a)**

*Exception: Diesel engines may be used in underground tunnels if the engines are permitted by DOSH.* **7069, 7070, 8470**

## Erection and Construction

Every year many workers lose their lives or are seriously injured when they fall or are crushed or struck because the structure they are erecting shifts or collapses. The following SOs address these hazards:

A. Truss and beam requirements

1. Trusses and beams must be braced laterally and progressively during construction to prevent buckling or overturning. The first member shall be plumb, connected, braced, or guyed against shifting before succeeding members are erected and secured to it. **1709(b)**

2. An erection plan and procedure must be provided for trusses and beams more than 25 ft. long. The plan must be prepared by a California-registered Professional Engineer, and it must be followed and kept available on the job site for inspection by Cal/OSHA staff. **1709(d)**

B. Structural steel erection **1710**

1. A load shall not be released from its hoisting line until the solid web structural members are secured at each connection with at least two bolts (of the same size/strength as indicated in the erection drawings), and drawn wrench tight. **1710(g)(1)**

2. Steel joists or steel joist girders shall not be placed on any support structure until the structure is stabilized. **1710(h)(1)(D)**

3. When steel joist(s) are landed on a structure, they shall be secured to prevent unintentional displacement prior to installation. **1710(h)(1)(E)**

4. Floors must be planked at every other story or 30 feet, whichever is less. **1635(b)(3), 1710 (l)(7)**

5. A floor must be installed within two floors below any tier of beams on which erection, riveting, bolting, welding, or painting is being done; otherwise, fall protection is required. **1635(b)(2)**

6. Fall protection is required when workers are connecting beams where the fall distance is greater than two stories or 30 ft., whichever is less. **1710(m)(1)**

   *Note: At heights over 15 feet and up to 30 feet, workers performing connecting must wear personal fall protection that gives them the ability to tie off.*

7. During work other than connecting operations, fall protection is required where the fall distance is greater than 15 ft. **1710(m)(2)**

8. Before any steel erection begins, the controlling contractor must provide the steel erector written notifications related to concrete strength and anchor bolt repair/replacement. **1710(c)**

9. Prior to removal of planking or metal decking, all employees must be instructed in the proper sequence of removal and safety. **1635(b)**

---

**Toolbox**

"Worker Deaths by Falls"

**PUB235**   www.oshatools.com

---

10. Requirements for the working area where floor openings are to be uncovered: **1635(c)**

   a) The area must be in the exclusive control of steel erection personnel and shall be barricaded to prohibit unauthorized entry.

   b) The floor area adjacent to the floor opening shall be barricaded or the floor opening shall be covered when not attended by steel erection personnel.

---

c) Floor openings shall be guarded by either temporary railings and toeboards or by covers. **1632(b)(1)**

d) Covers shall:

(1) Be capable of safely supporting the greater of 400 pounds or twice the weight of the employees, equipment and materials that may be imposed on any one square foot area of the cover at any time. **1632(b)(3)**

(2) Have not less than 12 in. of bearing on the surrounding structure. **1635(c)(3)**

(3) Be checked by a qualified person prior to each shift and following strong winds. **1635(c)(5)**

(4) Never be removed by walking forward where the walking surface cannot be seen. **1635(c)(6)**

(5) Bear a sign stating, "OPENING-DO NOT REMOVE", in 2 inch high, black bold letters on a yellow background. **1635(c)(4)**

11. Permanent Flooring -Skeleton Steel Construction in Tiered Buildings.

Unless the structural integrity is maintained by the design **1710(k)**:

a) There shall be not more than eight stories between the erection floor and the uppermost permanent floor.

b) There shall be not be more than four floors or 48 feet, whichever is less, of unfinished bolting or welding above the foundation or uppermost permanently secured floor.

12. All columns must be anchored by a minimum of 4 anchor bolts. **1710(f)(1)(A)**

*Exception: When columns are braced or guyed to provide the stability to support an eccentric load as specified in **1710(f)(1)(B)**.*

*Note: Persons engaged in steel erection should review and be knowledgeable of all the requirements contained in section **1710**.*

C. Wood/light gauge steel, residential and light commercial frame construction

1. Joists, beams, or girders of floors below the floor or level where work is being done, or about to be done, must be covered with flooring laid close together. **1635(a)(1)**

2. Employees shall not work from or walk on structural members until they are securely braced and supported. **1716.2(d)**

3. Before manually raising framed walls that are 15 ft. or more in height, temporary restraints, such as cleats on the foundation or floor system or straps on the wall bottom plate must be installed to prevent inadvertent horizontal sliding or uplift of the framed wall bottom plate. Anchor bolts alone shall not be used for blocking or bracing when raising framed walls 15 feet or more in height. **1716.2(c)**

4. When installing windows, wall openings shall be guarded as required by **1632**, however the guardrail may be removed for actual window installation if necessary. **1716.2(h)**

5. Scaffolds used as an edge protection platform must be fully planked, not more than 2 feet below the top plate, and located no more than 16 in. from the structure. **1716.2(i)(3)**

6. Employees exposed to fall hazards must be trained to recognize and minimize the fall hazard. **1716.2(j)**

7. Employees performing framing activities who are exposed to fall heights of 15 feet or greater must be protected by guardrails, personal fall protection systems or other effective means. **1716.2(e)**

# Ergonomics in Construction

Ergonomics is the study of improving the fit between the worker and the physical demands of the workplace. Ergonomics can be used to reduce injuries, improve productivity and reduce the costs of doing business.

The construction industry suffers from debilitating and costly occupational injuries primarily to workers' backs, necks, shoulders, hands and arms. These types of injuries or traumas are commonly called repetitive motion injuries (RMIs) and are caused by activities that are repeated on a regular basis. Symptoms of RMIs may include chronic pain, numbness, tingling, weakness and limited range of motion. RMI symptoms may not be noticeable for months or even years after exposures or may appear to be acute after a sudden and severe onset.

**Toolbox**

"Practical Demos of Ergonomic Principles"

**PUB229**   www.oshatools.com

A. Factors that can contribute to RMIs:

1. Awkward postures.

2. Forceful exertion, including heavy lifting.

3. Repetitive work.

4. Vibration from tools and equipment.

5. Pinching (contact stress) during tool use and material handling.

6. Temperature extremes.

7. Lack of recovery time to affected body parts.

   *Note: Repeated localized fatigue or soreness after completion of the same task or day's work often indicates that the worker is being exposed to conditions that can lead to RMIs.*

**Toolbox**

"Ergo Guidelines for Manual Material Handling"

**PUB250**   www.oshatools.com

B. Knowledge of ergonomic principles can be used to produce simple changes in the workplace and work activities which in turn can avoid injury, improve productivity, and make jobs easier. The requirements that employers must follow include: **5110**

1. Employers must establish and implement a program designed to minimize RMIs if more than one person is diagnosed with RMIs as follows:

   a) The RMIs are work related.

   b) The employees incurred the RMIs while performing a job process or operation of identical work activity.

   c) The RMIs were reported in the past 12 months.

   d) A licensed physician objectively identified and diagnosed the RMIs. **5110(a)**

   **Toolbox**
   "Ergonomic Guide for Carpenters and Framers"
   **PUB251**   www.oshatools.com

2. The program must include the following:

   a) A work site evaluation.

   b) Control of exposures that caused the RMIs.

   c) Training of employees. **5110(b)**

   **Toolbox**
   "Ergonomic Survival Guide for Laborers"
   **PUB254**   www.oshatools.com

C. Some ways to eliminate or reduce RMIs:

1. Proper lifting and material handling.

2. Use of equipment to reduce load and strain.

3. Employee rotation for repetitive tasks.

4. Use of ergonomically designed tools.

5. Use of personal protective equipment.

6. Appropriately timed rest periods.

## Excavation, Trenches, and Earthwork

Hazards associated with excavation are cave-ins; the striking of underground utilities; falling tools, materials, and equipment; and hazardous air contaminants or oxygen-deficient environments.

A. The minimum safety requirements are as follows:

1. Before opening an excavation these actions must be taken: **1541**

   a) Must identify subsurface installations prior to opening an excavation and ensure they are marked.

   b) Notify all regional notification centers and all subsurface installations owners who are not members of the notification centers, two working days before starting the work.

   *Exception: Emergency repair work to subsurface facilities done in response to an emergency as defined in Government Code Section **4216(d)**.*

   c) Must receive a positive response from all known owners/operators of subsurface installations.

   d) Must meet with owners/operators of high priority (such as high pressure pipelines, natural gas/ petroleum pipelines, electrical lines greater than 60,000 volts etc.) subsurface installations, that are located within 10 feet of the proposed excavation.

   e) Only qualified persons (persons that meet training and competency requirements) can perform subsurface installation locating activities.

   f) All exposed employees must be trained in excavator notification/excavation activities.

   g) Obtain a permit from DOSH if workers are required to enter an excavation that is 5 ft. deep or deeper. **341(a)(1)**

2. While excavating, the exact locations of the underground utilities must be determined by safe and acceptable means. **1541(b)(3)**

> **Toolbox**
>
> "Preventing Worker Deaths in Trenches"
>
> **PUB231**   www.oshatools.com

3. Excavators must immediately notify the subsurface installation owner/operator of any damage discovered during or caused by excavating activities. If the damage or escaping material endangers life or property, immediately notify 911.

4. While the excavation is open, the underground utilities must be protected, supported, or removed as necessary. **1541(b)(4)**

> **Toolbox**
>
> "Trench and Excavation"
>
> **PUB216**   www.oshatools.com

B. When employees are in an excavation, the following requirements apply:

1. Employees shall be protected from cave-ins by an appropriate protective system. **1541.1(a)(1)**

   *Exception: If excavations are made entirely in stable rock, or are less than 5 ft. deep, and a competent person has determined that there is no potential for a cave-in, no protective system is needed.*

2. A competent person must be on site to do the following:

   a) Conduct inspections of the excavations, adjacent areas, and protective systems before the start of work; as needed throughout the shift; and daily for potential cave ins, failures, hazardous atmospheres, or other hazards. **1541(k)(1)**

   b) Take prompt corrective action or remove employees from the hazard.

3. The competent person must be able to demonstrate the following:

a) The ability to recognize all possible hazards associated with excavation work and to test for hazardous atmospheres.

b) Knowledge of the current safety orders pertaining to excavation and trenching.

c) The ability to analyze and classify soils.

d) Knowledge of the design and use of protective systems.

e) The authority and ability to take prompt corrective action when conditions change.

C. Requirements for protective systems include the following:

1. Protective system design must be based on soil classification: Stable rock, Type A, B, or C soils. **1541.1 Appendix A (b), (c)**

2. Soil classification is required as follows unless the protective system design is based on Type C soil:

a) Classification must take into account both site and environmental conditions. **1541.1 Appendix A (a)(1)**

b) Soil must be classified by a competent person as Type A, B, or C soil. **1541.1 Appendix A (c)(1)**

c) Classification must be based on the results of at least one visual and at least one manual analysis. **1541.1 Appendix A (c)(2)**

## Table I

### Site Conditions That Affect Rock/Soil Slope Stability

| Condition | Requirement |
|---|---|
| Soil is fractured/unstable dry rock. | Downgrade to Type B. |
| Soil is fractured/unstable submerged rock. | Downgrade to Type C. |
| Soil is cemented (caliche, hardpan, etc.). | Classify as Type A. |
| Soil is fissured. | Downgrade from Type A to Type B. |
| Soil is subject to vibration. | Downgrade from Type A to Type B. |
| Soil has been previously disturbed. | Downgrade from Type A to Type B. |
| Soil is submerged or water is freely seeping through the sides of the excavation. | Downgrade from Type A to Type C.<br><br>Downgrade from Type B to Type C. |
| Soil profile is layered with the layers dipping into the excavation on a slope of four horizontal to one vertical or steeper. | Downgrade from Type A to Type C.<br><br>Downgrade from Type B to Type C. |

# Illustration 5

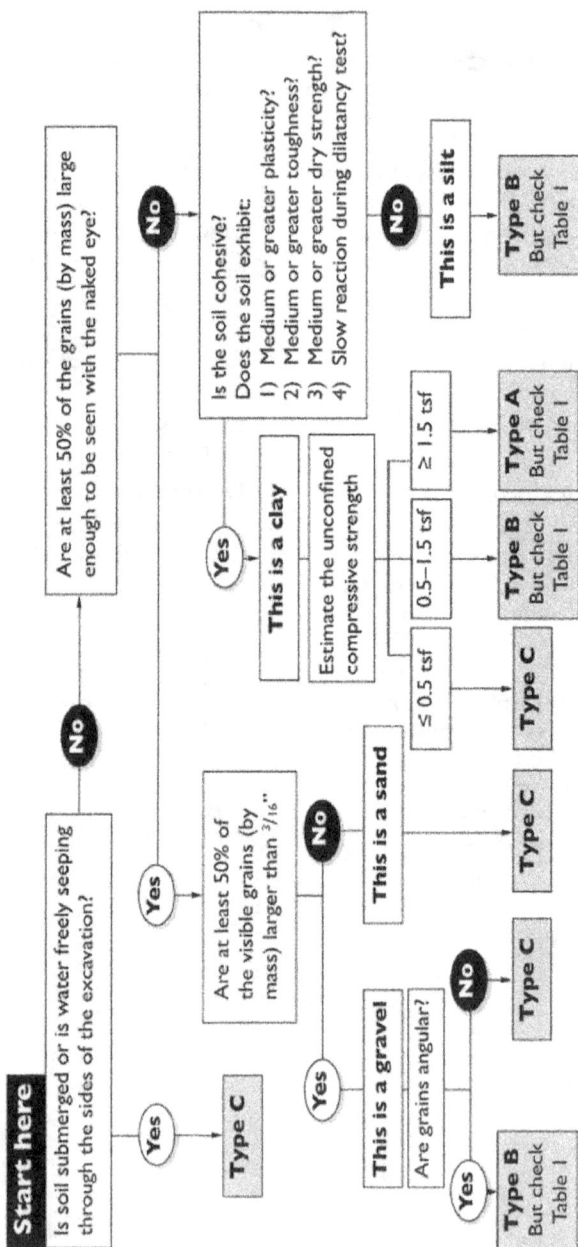

**Start here**

Is soil submerged or is water freely seeping through the sides of the excavation?

→ **Yes** → **Type C**

→ **No** →

Are at least 50% of the grains (by mass) large enough to be seen with the naked eye?

**No** →

Is the soil cohesive?
Does the soil exhibit:
1) Medium or greater plasticity?
2) Medium or greater toughness?
3) Medium or greater dry strength?
4) Slow reaction during dilatancy test?

**Yes** → **This is a clay**

Estimate the unconfined compressive strength

- ≤ 0.5 tsf → **Type C**
- 0.5–1.5 tsf → **Type B** But check Table I
- ≥ 1.5 tsf → **Type A** But check Table I

**No** → **This is a silt** → **Type B** But check Table I

**Yes** →

Are at least 50% of the visible grains (by mass) larger than 3/16"?

**No** → **This is a sand** → **Type C**

**Yes** → **This is a gravel**

Are grains angular?

**No** → **Type C**

**Yes** → **Type B** But check Table I

3. Standard shoring, sloping, and benching must be used as specified in **1540** and **1541.1(b)** or according to tabulated data prepared by a registered engineer (see illustrations 6-8 below).

4. Protective systems for excavations deeper than 20 ft. shall be designed by a registered engineer. **1541.1 Appendix F**

5. Additional bracing must be used when vibration or surcharge loads are a hazard. **1541.1 Appendix A**

6. Excavations must be inspected as needed after every rainstorm, earthquake, or other hazard increasing occurrence. (Water in the excavation may require a reclassification of soil type). **1541(k)(1)**

7. Employees must be protected from falling materials by scaling, installation of protective barriers, or other methods. **1541(j)(1)**

8. Uprights shall extend to the top of the trench and its lower end not more than 2 feet from the bottom of the trench. **1541(j)(1)**

9. Employees must be protected from excavated or other material by keeping such material 2 ft. from the excavation edge or by using barrier devices. **1541(j)(2)**

10. Ladders or other safe access must be provided within 25 ft. of a work area in trenches 4 ft. or deeper. **1541(c)(2)**

11. Excavation beneath the level of adjacent foundations, retaining walls, or other structures is prohibited unless requirements of **1541(i)** have been met. **1541(i)(1)**

12. Shored, braced, or underpinned structures must be inspected daily when stability is in danger. **1541(i)(2)**

# Illustration 6

BENCHING & SLOPING FOR EXCAVATIONS MADE IN TYPE "A" SOIL

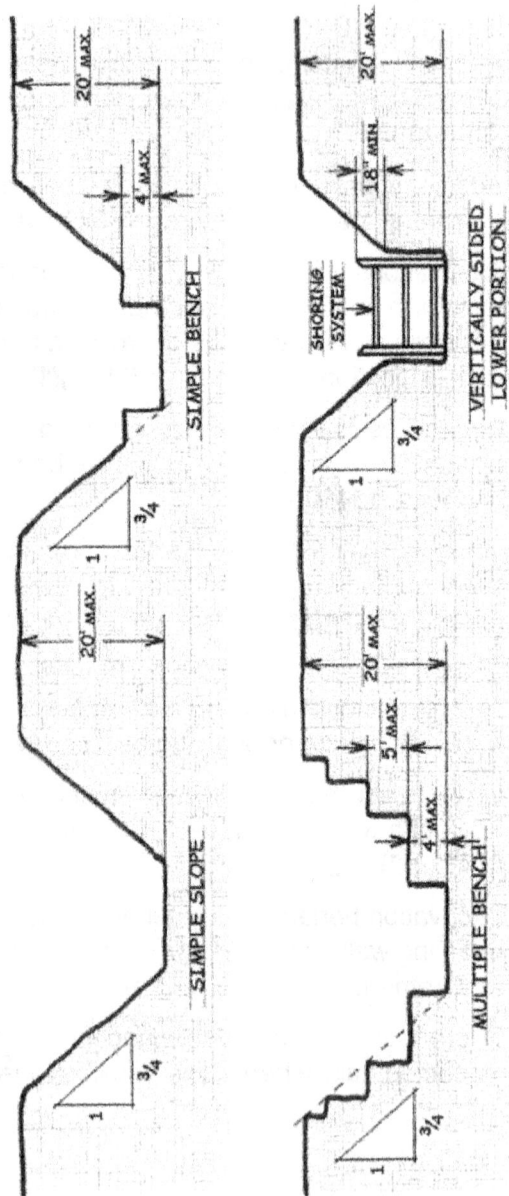

SIMPLE SLOPE

SIMPLE BENCH

MULTIPLE BENCH

VERTICALLY SIDED LOWER PORTION

# Illustration 7

## BENCHING & SLOPING FOR EXCAVATIONS MADE IN TYPE "B" SOIL

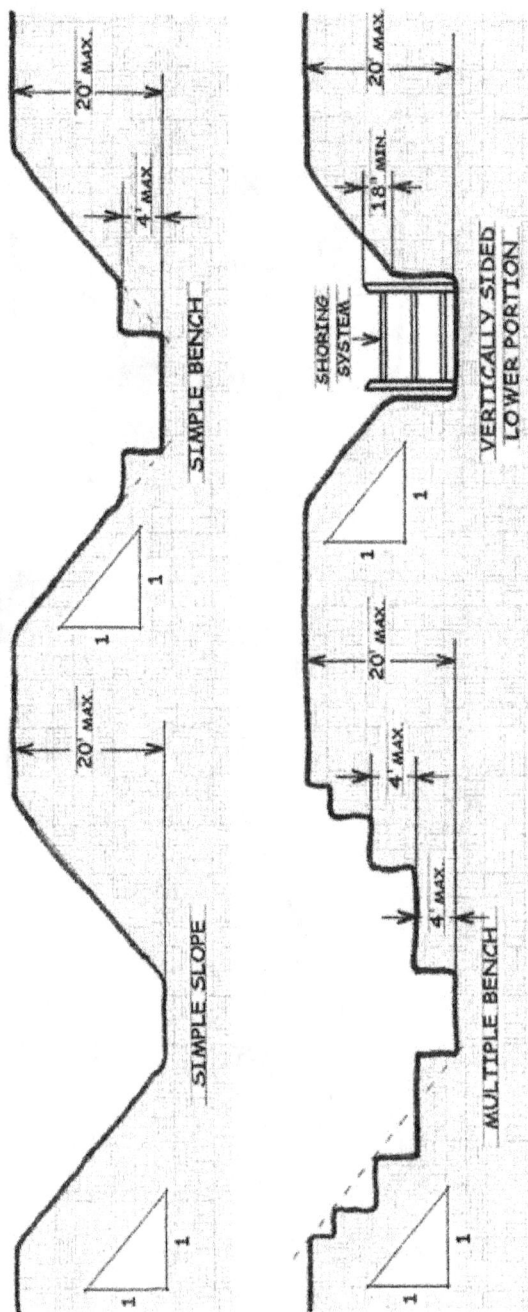

SIMPLE SLOPE

SIMPLE BENCH

MULTIPLE BENCH

VERTICALLY SIDED
LOWER PORTION

SHORING
SYSTEM

20' MAX.

4' MAX.

18" MIN.

1
1

# Illustration 8

BENCHING & SLOPING FOR EXCAVATIONS MADE IN TYPE "C" SOIL

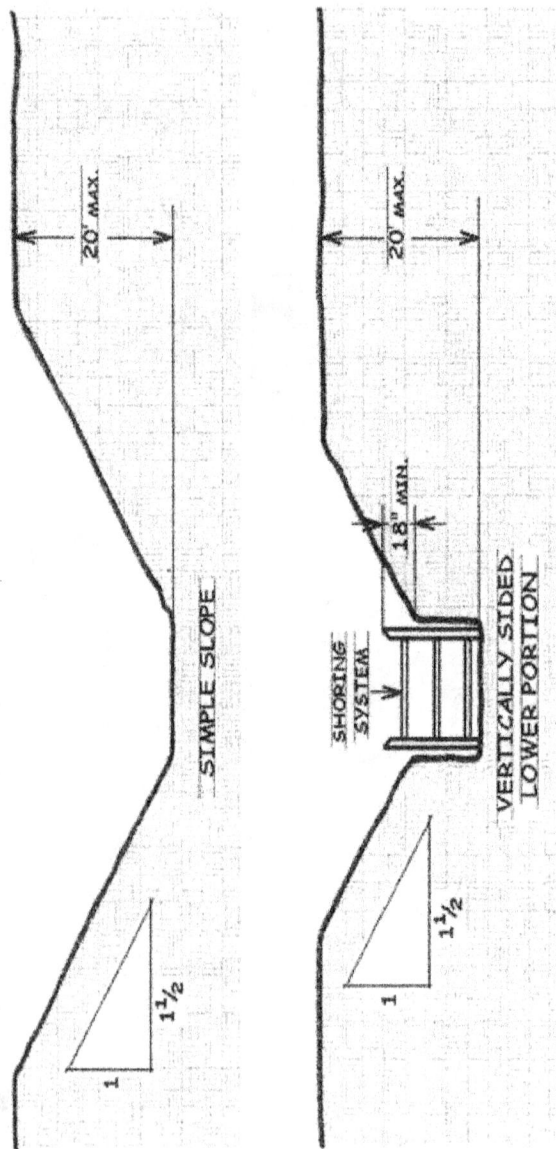

SIMPLE SLOPE

20' MAX.

1½
1

VERTICALLY SIDED
LOWER PORTION

SHORING SYSTEM

20' MAX.

18" MIN.

1½
1

13. Walkways or bridges with standard guardrails must be installed when employees or equipment are required or permitted to cross over excavations that are at least 6 ft. deep and wider than 30 in. **1541(l)(1)**

14. Barriers must be erected around excavations in remote locations. All wells, pits, shafts, and caissons must be covered or barricaded, or if temporary, backfilled when work is completed. **1541(l)(2)**

D. Safety orders pertaining to shafts and wells include the following:

1. All shafts and wells more than 5 ft. deep into which workers are required to enter must be retained with lagging, spiling, or casing. **1542(a)(1)**

2. Tests or procedures shall be performed before entry into exploration shafts to ensure the absence of dangerous air contamination or oxygen deficiency. **1542(c)(3), 5158**

3. An employee entering a bell-bottom pier hole or other deep or confined-footing excavation shall wear a harness that has a lifeline attended by another employee. **1541(g)(2)(B)**

4. Shafts in other than hard, compact soil shall be completely lagged and braced. **1542(c)(1)**

5. Head protection is required for workers who enter a well or shaft. **3381**

6. Shafts more than 20 ft. deep are subject to the TSOs. **8403(a)**

## Explosion Hazards

At times employees may be exposed to explosion hazards without their knowledge. In addition to substances (such as dynamite) that are designed specifically for the purpose of creating explosions, there are substances that will cause an explosion when present in certain concentrations and exposed to an ignition source. SOs to control these hazards include:

A. Combustible dust:

1. Combustible dust concentrations must be controlled at or below 25% of the lower explosive limit (LEL) unless all ignition sources are eliminated or identified and specifically controlled. **5174(a)**

2. Accumulated and settled combustible dusts must be cleaned up to prevent a fire or explosion. **5174(b)**

3. Cleaning with compressed air and blowing combustible dust may be done only when other methods cannot be used, when possible sources of ignition have been eliminated, and when hoses and nozzles are grounded. **5174(f)**

B. Flammable vapors:

1. Ventilation in enclosed places must be sufficient to prevent flammable vapor or gas concentrations from exceeding 25% of the LEL. **5416(a)**

2. No source of ignition is permitted indoors or outdoors where vapor or gas concentrations may reasonably be expected to exceed 25% of the LEL. **5416(c)**

3. Employers need to be aware that most flammable vapors are toxic even at a very low concentration and can cause adverse health effects. Employers must have control measures to keep employees safe.

   *Note: Check also for confined space conditions and hazardous locations. **5158**, **2540.1**, Confined Spaces section of this guide.*

## Fall Protection

T8 CCR includes fall protection standards in various sections of the GISOs, CSOs, TSOs, and ESOs. These standards reflect the levels of the fall hazards associated with each activity.

A. The factors affecting the level of hazard include the following:

1. Fall height.

2. Level of hazard awareness and skill of the employee.

3. Physical work environment (e.g., conditions affecting the stability of the employee on the work surface).

4. Duration of exposure to the fall hazard.

*Note: Because factors 2, 3, and 4 listed above vary with different trades and activities, the regulatory requirements for fall protection reflect those differences. Below find definitions and selected fall protection requirements:*

> **Toolbox**
>
> "Worker Deaths by Falls"
>
> **PUB235**   www.oshatools.com

B. A personal fall protection (PFP) system prevents a worker from falling or—if the worker is falling—stops the fall. PFP systems include guardrails, safety nets, personal fall restraint systems, personal fall arrest systems, and positioning device systems.

1. Guardrails are required to guard the open sides of all work surfaces that are 7 1/2 ft. or higher or workers must be protected by other means. The railing must be made from select lumber (Doug fir#1 or better 1500 Psi or equivalent) and must consist of a top rail 42 in. to 45 in. high, 2" x 4" (min.); a 1" x 6" mid rail halfway between the top rail and the floor; and support posts at least 2" x 4" at 8 ft. o.c.

2. A personal fall restraint (PFR) system is used to prevent an employee from falling. It consists of anchorages, connectors, and a body belt or harness. It may include lanyards, lifelines, and rope grabs designed for that purpose.

3. A personal fall arrest (PFA) system is used to stop an employee during a fall from a working level and to keep him or her from hitting a lower level or structure. The system consists of an anchorage, connectors, and a body harness. It may include a lanyard, a lifeline, a deceleration device, or suitable combinations of these. A PFA system must meet the following requirements:

   a) It must limit the maximum arresting force on an employee to 1,800 lbs.

   b) It must be rigged such that an employee can neither free fall more than 6 feet, nor contact any lower level, and, where practicable, the anchor end of the lanyard shall be secured at a level not lower than the employee's waist.

   c) Anchorage points must be able to support 5,000 lbs. per employee attached or:

      (1) Must be designed, installed, and used as part of a complete PFA system with a safety factor of two; and

      (2) Under the supervision of a qualified person.

   d) The PFA system lifeline must meet the following requirements: **1670(b)**

      (1) It must be able to support 5,000 lbs.

      (2) Each employee must be attached to a separate lifeline. **1670(b)(4)**

> **Toolbox**
> "Preventing Falls From Communication Towers"
> **PUB230**   www.oshatools.com

*Exception: During the construction of elevator shafts, two employees may be attached to a lifeline that is able to support 10,000 lbs.*

(3) The lower end of the vertical lifeline must extend to within 4 ft. from the ground. **1504**

(4) A horizontal lifeline system must be designed, installed, and used under the supervision of a qualified person and maintained with a safety factor of at least two. **1670(b)(2)**

*Note: The use of a body belt as a part of a PFA system is prohibited.* **1670(b)**

4. Body belts, harnesses, and components shall be used only for employee protection and not to hoist materials. Body belts used in conjunction with fall restraint systems or positioning devices shall limit the maximum arresting force on an employee to 900 pounds. **1670(b)**

5. Safety nets may be used in place of all other fall protection systems if the nets are installed properly. **1671**

C. A PFP system must be used if guard railing or safety nets are not installed for the following fall distances and work activities:

1. A fall distance of more than 6 ft., when placing or tying rebar in walls, columns, piers, etc. **1712(e)**

*Exception: A PFP system is not required during point-to-point horizontal or vertical travel on rebar up to 24 feet above the surface below if there are no impalement hazards.* **1712(e)**

2. A fall distance of 7 1/2 ft. or greater during the following:

a) Work from the perimeter of a structure, through shaft-ways and openings. **1670(a)**

b) Work anywhere on roofs with slopes greater than 7:12. **1670(a)**

c) Work from thrust-outs or similar locations when the worker's footing is less than 3 1/2 in. wide. **1669(a)**

d) Work on suspended staging, floats, catwalks, walkways, or advertising sign platforms. **1670(a)**

e) Work from slopes steeper than 40 degrees. **1670(a)**

3. A fall distance of 15 ft. or greater during the following:

a) Work from buildings, bridges, structures on construction members, such as trusses, beams, purlins, or plates that are of at least 4" nominal width. **1669(a)**

b) Ironwork other than connecting. **1710(g)(2)**

c) Work on structural wood framing systems and during framing activities on wood or light gauge steel frame residential/light commercial construction. **1716.1(c)(1), 1716.2(e)**

*Exception: For residential/light commercial frame construction, workers are considered protected when working on braced joists, rafters or roof trusses spaced on 24 inch (or less) centers when they work more than 6 feet from unprotected sides or edges.*

4. An eave height of 20 ft. or greater, during all roofing operations (see exceptions in 2a above and 6a and 6b below). **1730(b)**

5. A fall distance of 30 ft. or greater, when ironworkers are connecting structural beams. **1710(g)(1)**

6. Any height during work:

a) On roofs sloped steeper than 7:12 the air hose for the pneumatic nailer shall be secured at roof level in such a manner as to provide ample, but not excessive, amounts of hose. **1704(d)**

b) On roofs, while an operator uses a felt-laying machine or other equipment that requires the operator to walk backwards (see prohibitions). **1730(d)**

c) From boatswain's chairs. **1662(c)**

d) From float scaffolds. **1663(a)(5)**

e) From needle-beam scaffolds. **1664(a)(12)**

f) From suspended scaffolds. **1660(g)**

D. A fall protection plan (FPP) must be implemented when a fall protection (FP) system is required but cannot be used because the system creates a greater hazard or is impractical. **1671.1** The fall protection plan must: **1671.1(a)(1)**

1. Be prepared by a qualified person (QP) who is identified in the plan.

2. Be developed for a specific site or developed for essentially identical operations.

3. Be updated by the QP.

4. Document why a conventional FP system cannot be used.

5. Identify the competent person to implement and supervise the FPP.

6. Identify the controlled access zone for each location where a conventional FP system cannot be used.

7. Identify employees allowed in the CAZ.

8. Be implemented and supervised by the competent person.

*Note: An up-to-date copy of the fall protection plan must be at the job site.*

E. The controlled access zone must be established and maintained as follows: **1671.2**

  1. A control line or its equivalent must control access to the CAZ and must:

     a) Consist of ropes, wires, tapes, or equivalent materials and be supported by stanchions.

     b) Be flagged or marked at not more than 6 ft. o.c.

     c) Be rigged not fewer than 39 in. and not more than 45 in. from the working surface.

     d) Have a breaking strength of 200 lbs. (min.). See **1671.2** for greater detail.

  2. Signs must be posted to keep out unauthorized persons.

  3. A safety monitoring system is required and must include a designated safety monitor who is able to:

     a) Monitor the safety of other employees.

     b) Recognize fall hazards.

     c) Warn an employee when it appears that the employee is unaware of a fall hazard or is acting in an unsafe manner.

     d) Stay in sight of and in communication with the employee being monitored.

     e) Have no other responsibilities. **1671.2**

  *Note: Only an employee covered by a fall protection plan shall be allowed in a CAZ.*

F. Fall protection for production type residential roofing work: **1731(c)**

  1. For Roof Slopes 3:12 through 7:12, the following applies: Employees shall be protected from falling when on a roof surface where the eave height exceeds 15 feet above the grade or level below by use of one or any combination of the following methods:

a) Personal Fall Protection. **1670**

b) Catch Platforms. **1724**

c) Scaffold Platforms. **1724**

d) Eave Barriers. **1724**

e) Standard Railings and Toeboards. **1620, 1621**

f) Roof Jack Systems. **1724**

2. For Roof Slopes Steeper than 7:12, the following applies: Employees shall be protected regardless of height from falling by methods prescribed above with exception of Eave Barriers and Roof Jack Systems.

G. Section **1730** applies to all roofing work that are not on new production-type residential construction with roof slopes 3:12 or greater. **1730(f)(6)**

## Fire Protection and Prevention

The employer is responsible for establishing an effective fire prevention program and ensuring that it is followed throughout all phases of the construction work. **1920(a)**

A. Fire-fighting equipment must be:

1. Freely accessible at all times. **1920(b)**

2. Placed in a conspicuous location. **1920(c)**

3. Well maintained. **1920(d)**

B. A water supply that is adequate to operate fire-fighting equipment must be made available as soon as combustible materials accumulate. **1921(a)**

C. Fire extinguisher use must comply with the following:

1. Fire extinguishers must be kept fully charged, inspected monthly, and serviced annually. **1922(a)**

2. At least one fire extinguisher, rated not less than 2A, must be provided at each floor.

3. At least one fire extinguisher, rated not less than 2A, must be provided adjacent to the stairway at each floor level.

4. Fire extinguishers rated not less than 2A must be provided for each 3,000 ft. of floor area or a fraction thereof.

5. Fire extinguishers must be kept within 75 ft. of the protected area. **1922(a)**

*Exception: Fire extinguishers must be kept within 50 ft. of wherever more than 5 gal. of flammable or combustible liquid or 5 lbs. of flammable gas is being used. 1922(a)*

6. Training in the use of fire extinguishers must be provided annually. **6151(g)**

*Note: See specific SOs and manufacturing specifications for appropriate use of fire extinguishers.*

## First Aid

Regulations concerning first aid include the following:

A. A first aid kit must be provided by each employer on all job sites and must contain the minimum of supplies as determined by an authorized licensed physician or as listed in **1512(c)**.

B. Trained personnel in possession of a current Red Cross First Aid certificate or its equivalent must be immediately available at the job site to provide first aid treatment. **1504(a), 1512(b)**

C. Each employer shall inform all of its employees of the procedure to follow in case of injury or illness. **1512(d)**

D. Emergency medical services, including a written plan, must be provided. **1512(a), (e)**

E. Exposure to blood borne pathogens is considered a job-related hazard for construction workers who are assigned first aid duties in addition to construction work. Although construction employers are specifically exempted from **GISO 5193** requirements, they are required to provide appropriate protection for employees who may be exposed to blood borne pathogens when providing first aid. **3203**

# Flaggers

Flaggers must be used at locations on a construction site as soon as barricades and warning signs cannot effectively control moving traffic. The employer must ensure the following:

A. Flaggers must be placed in locations so as to give effective warning. **1599(b)**

B. Worksite traffic controls and placement of warning signs must now conform to the requirements of the "California Manual on Uniform Traffic Control Devices for Streets and Highways" (the Manual), published by CalTrans. **1598(a), 1599(c)**

C. Warning signs must be placed according to the "the Manual." **1599(c)**

> ### Toolbox
> "California Manual on Uniform Traffic Control Devices, Temporary Traffic Control, Part 1&6"
> **PUB247**   www.oshatools.com

D. Flaggers must wear orange or strong yellow-green warning garments, such as vests, jackets, shirts, or rainwear. **1599(d)**

> ### Toolbox
> "Work Zone Operations, Best Practices"
> **PUB246**   www.oshatools.com

E. The employer shall select the proper type (class) of high visibility safety apparel for a given occupational activity by consulting the Manual, apparel manufacturer, ANSI/ ISEA 107-2004, Appendix B or the American Traffic Safety Services Association (ATSSA). **1599(f)**

F. Flaggers shall wear warning garments manufactured in accordance with the requirements of ANSI)/ ISEA 107-2004, High Visibility Safety Apparel and Headwear. **1599(d)**

G. During the hours of darkness: **1599(e)**

» The flagger shall be clearly visible to approaching traffic and be outfitted with reflectorized garments manufactured in accordance with the requirements of the ANSI/ ISEA 107-2004, High Visibility Safety Apparel and Headwear.

» The retroreflective material shall be visible at a minimum distance of 1,000 feet.

» During snow or fog conditions, only colored vests, jackets and/or shirts with retroreflective material that meets the ANSI/ISEA and the minimum distance requirements shall be worn.

H. Flaggers must be trained. **1599(g)**

I. Training must be documented in accordance with the IIP Program requirements. **1599(g)**

---

**Toolbox**
"Building Safer Highway Work Zones"
**PUB218**   www.oshatools.com

---

## Flammable and Combustible Liquids

Flammable and combustible liquids include gasoline, paint thinners, solvents, etc.

A. These liquids must be kept in closed containers when not in use. **1935(a)**

B. Leakage or spillage must be disposed of promptly and safely. **1935(b)**

---

C. Flammable and combustible liquids may be used only where no open flames or sources of ignition exist (see specifics in **1935(c)**).

D. All containers of flammable and combustible liquids must be plainly marked with a warning legend. **5417(a)**

E. Flammable liquids must not be used: **5417**

   » To wash floors, structures, or equipment except where there is adequate ventilation.

   » To spray for cleaning purposes unless the liquids are used in a spray booth or outdoors where there is no ignition source within 25 ft. of their use.

F. Flammable liquids must be stored and transported in closed containers. **5417(e)**

*Note: For specific requirements concerning indoor and outdoor storage, see **1931** and **1932**. For on-site dispensing operations see **1934**.*

G. A hazard communication program is required. **5194**

> **Toolbox**
> "Pocket Guide to Chemical Hazards"
> **PUB228**  www.oshatools.com

## Forklifts

Safety regulations concerning the use of forklifts are as follows:

A. Industrial trucks such as forklifts shall be designed, constructed, and maintained in accordance with the applicable standards. **3650(c)**

B. The employer shall establish and enforce a system to prevent trucks, trailers or railcars from pulling away from the loading dock before the loading or unloading operation is completed. Trucks, trailers, and railcars boarded by fork lifts during loading dock operations shall be secured against unintended movement. **3336**

C. The rated lifting capacity of the forklift must be posted in a location readily visible to the operator. **3660(a)**

D. Elevating employees requires the following:

1. The forklift must be equipped with a platform not less than 24" x 24" in size.

   a) The platform must be properly secured to the forks or the mast.

   b) The platform must be equipped with guardrails, toe boards, and a back guard.

   c) It must have no spaces or holes larger than 1 in.

   d) It must have a slip-resistant platform surface. **3657(a)(2)**

2. The operator must be at the controls while the employees are elevated. **3657(d)**

3. The operator must be instructed in the operating rules for elevating employees. **3657(i)**

4. Employees shall not sit, climb, or stand on platform guardrails or use planks, ladders, or other devices to gain elevation. **3657(h)**

*Note: When guardrails are not possible, personal fall protection is required. 3657(b)*

E. All forklifts must have parking brakes. **3661(b)**

F. All forklifts must have an operable horn. **3661(c)**

G. When the operator is exposed to the possibility of falling objects, the forklift must be equipped with overhead protection (canopy). **3657(c)**

H. When provided by the industrial truck manufacturer, an operator restraint system such as a seat belt shall be used. **3650(t)**

I. Seat belt assemblies shall be provided and used on all equipment where rollover protection is installed. **3653(a)**

J. The employer must post and enforce a set of operating rules that include the following: **3650(s)**

1. Only trained and authorized drivers may operate forklifts.

2. Stunt driving and horseplay are prohibited.

3. Employees must not ride on the forks.

4. Employees must never be permitted under the forks (unless forks are blocked).

5. The driver must inspect the vehicle once during a shift.

6. The operator must look in the direction of travel and must not move the vehicle until all persons are clear of the vehicle.

7. Forks must be carried as low as possible.

8. The operator must lower the forks, shut off the engine, and set the brakes (or block the wheels) before leaving the forklift unattended (that is, when the operator is out of sight of the vehicle or 25 ft. away from it).

9. Trucks must be blocked and brakes must be set before a forklift is driven onto the truck bed.

10. Extreme care must be taken when tilting elevated loads.

11. The forklift must have operable brakes capable of stopping it safely when it is fully loaded.

K. An employee must be properly trained (as certified by the employer) before operating a forklift. **3668(a)**

1. An evaluation of the operator's performance must be conducted at least once every three years. **3668(d)**

2. Refresher training in relevant topics must be provided to the operator when: **3668(d)(1)**

    a) The operator is observed operating the vehicle in an unsafe manner.

    b) The operator has been involved in an accident or near-miss incident.

c) The operator's evaluation reveals that he or she is not operating the truck safely.

d) The operator is assigned to drive a different type of truck.

e) Changes in workplace conditions could affect safe operation of the truck.

## Forms, Falsework, and Vertical Shoring

By definition concrete forms are considered falsework. Falsework, however, also includes support systems for forms, newly completed floors, bridge spans, etc., that provide support until appropriate curing or stressing processes have been completed.

See below for selected SOs:

A. Design of falsework

1. Concrete formwork and falsework must be designed, supported, and braced to safely withstand the intended load. **1717(a)(1)**

2. Falsework design, detailed calculations, and drawings must be signed and approved by an engineer (Ca PE) if the falsework height (sill to soffit) exceeds 14 ft., if the individual horizontal span length exceeds 16 ft., or if vehicle or railroad traffic goes through the falsework. **1717(b)(1)(A), (B)**

   *Note: For other falsework, approval may be provided by a manufacturer's representative or a licensed contractor's qualified representative. **1717(b)(2)(B), (C)***

3. Falsework plans must be available at the job site. **1717(b)(3)**

4. Minimum design loads are as follows: **1717(a)(2)**

   a) Total combined live and dead load: 100 psf

   b) Live load and formwork: 20 psf

---

5. Additional loads must be considered in the design. **1717(a)**

B. Erection of falsework

1. Falsework must be erected on a stable, level, compacted base and supported by adequate pads, plates, or sills. **1717(b)(4)**

2. Shore clamps (metal) must be installed in accord with manufacturer's instructions. **1717(d)(2)**

C. Inspection

1. Before pouring concrete on falsework requiring design approval, an engineer (CaPE) or the engineer's representative must inspect for and certify compliance with plans. **1717(c)(1)**

   *Note: For other falsework, the inspection and certification may be provided by a manufacturer's representative or a licensed contractor's qualified representative. 1717(c)(2)(B), (C)*

2. A copy of the inspection certification must be available at the job site. **1717(c)(3)**

D. Access to forms and falsework

1. Joists (5 1/2 in. wide) at not more than 36 in. o.c. may be used as walkways while forms are placed. **1717(d)(3)**

2. A plank (12 in. wide) may be used as a walkway while joists are placed. **1717(d)(5)**

E. Fall protection

Periphery rails are required as soon as supporting members are in place. **1717(d)(4)**

*Note: The area under formwork is a restricted area and must be posted with perimeter warning signs. 1717(d)(6)(A)*

# Guardrails

Guardrails must be installed at the open sides of all work surfaces that are 7 1/2 ft. or higher above the ground, floor, or level underneath, or workers must be protected by other fall protection or, if justified, by a valid fall protection plan. **1621(a)**

A. Guardrailing specifications. **1620**

1. Railings shall be constructed of wood or in an equally substantial manner from other materials, and shall consist of the following:

   a) A wooden top rail that is 42 in. to 45 in. high and that measures 2 in. x 4 in. or larger.

   b) A mid-rail shall measure at least 1 in. x 6 in., and shall be placed halfway between the top rail and the floor when there is no wall or the parapet wall is less than 21 in. high.

   c) Screens, mesh, intermediate vertical members, solid panels or equivalent members, may be used in lieu of a mid-rail subject to the following:

      (1) Screens and mesh shall extend from the top rail to the floor and along the entire opening between top rail supports.

      (2) The gap between the intermediate vertical members shall not be greater than 19 in.

      (3) Other intermediate members such as solid panels shall not have gaps more than 19 in.

   d) Wood posts shall be not less than 2 in. by 4 in. in cross section, spaced at 8-foot or closer intervals.

   *Notes: Use only "Selected lumber" - free from damage that affects its strength for wood railings. Steel banding and plastic banding shall not be used as top rails or mid-rails.*

2. All railings and components shall be capable of withstanding a force of at least 200 pounds applied to the top rail within 2 in. of the top edge, in any outward or downward direction, at any point along the top edge.

3. Mid-rails, screens, mesh, intermediate vertical members, solid panels, and equivalent members shall be capable of withstanding a force of at least 150 pounds applied in any downward or outward direction at any point.

4. The top rail or midrail on scaffolding platforms may be substituted by the X-braces (see the Scaffolds section in this guide). **1644(a)(6)**

5. The ends of the rails shall not overhang the terminal posts, except where such overhang does not constitute a projection hazard. **1620(f)**

6. Railings shall be so surfaced as to prevent injury to an employee from punctures or lacerations, and to prevent snagging of clothing. **1620(g)**

B. Guardrailing applications

1. Floor and roof openings: **1632(b)(3)**

   a) Floor, roof, and skylights openings in any work surface must be guarded by railings and toeboards or by covers.

   b) The cover must be able to support 400 pounds or twice the weight of the employees, equipment, and material, and be securely fastened.

   c) Covers must bear a sign, with minimum 1 inch letters, stating - OPENING - DO NOT REMOVE.

   d) Employees within 6 feet of any skylight shall be protected from falling through the skylight opening by any one of the following methods:

   (1) Guardrails. **3209**

   (2) Skylight screens. **3212**

   (3) Personal fall protection system. **1670**

   (4) Covers installed over the skylights. **1632**

   (5) Fall protection plan. **1671.1**

   *Exception: When the work is of short duration and exposure is limited.* **3212(e)**

e) Access to surfaces glazed with transparent or translucent materials are not permitted unless an engineer certifies that the surface will sustain all anticipated loads. **3212(f)**

2. Wall openings: Wall openings must be guarded if there is a drop of more than 4 ft. and the bottom of the opening is less than 3 ft. above the working surface. **1632(j)**

3. Elevators: Guardrails are required for elevator shaft openings that are not enclosed or do not have cages. **1633**

4. Falsework: Guardrails are required as soon as falsework-supporting members are in place. **1717(d)(4)**

5. Demolition: Wall openings must be guard-railed during demolition except on the floor being demolished and on the ground floor. **1735(k)**

6. Roofing operations: Provisions must be made during roofing operations to prevent workers from falling off roofs 20 ft. or higher. **1730(b)(1)**

7. Skeleton steel building: A single 3/8-in. wire rope, in lieu of standard railing, may be used to guard openings and exposed edges of temporary floors or planking in skeleton steel buildings. The 3/8-in. wire rope must have a breaking strength of 13,500 lbs. (min.) and be placed at 42 in. to 45 in. above the finished floor. **1710(l)(3)**

# Hazard Communication Program (Haz-Com)

Hazardous substances are generally defined as substances likely to cause injury or illness because they are explosive, flammable, toxic, poisonous, corrosive, oxidizing, irritant, or otherwise harmful. These substances may include solvents, paints, thinners, cleaning agents, fresh concrete, and fuels. Employers whose employees may be exposed to hazardous substances are required to have a Haz-Com program. **5194**

A. The program must include the following:

1. A list of the hazardous substances that are used or stored in the workplace. Hazardous substances that require a haz-com program are:

   a) Any substance that is a physical or a health hazard.

   b) Any hazardous substance listed in the following:

      (1) The Hazardous Substances List. **339**

      (2) The Code of Federal Regulations (CFR, Part 1910, Subpart Z).

      (3) Threshold Limit Values for Chemical Substances in the Work Environment (ACGIH) 1991 - 1992.

      (4) Regulated Carcinogens. **5209**

      (5) Eleventh Report on Carcinogens, National Toxicology Program, 2005.

      (6) Monographs, International Agency for Research on Cancer, Volumes 1 - 53, and Supplements 1 - 8, World Health Organization.

      (7) Material Safety Data Sheets (MSDSs) on reproductive toxicants or cancer-producing substances.

      (8) T22 CCR **12000** (Proposition 65).

---

**Toolbox**

"Pocket Guide to Chemical Hazards"

**PUB228**  www.oshatools.com

---

2. Labels and other forms of warning on containers of hazardous substances.

3. Readily accessible MSDSs.

4. Procedures for safe handling, use, storage, disposal, and clean-up to protect employees.

5. Training on the hazardous substances that employees are or could be exposed to in the workplace.

6. A plan for managing multi-employer work-site issues.

7. A plan for periodically (e.g., annually) evaluating the effectiveness of the program and for updating the program.

B. The Haz-Com program must be in writing and must be available on request to employees, their representatives, and Cal/OSHA.

---

**Toolbox**

"California Hazard Communication Regulation"

**PUB257**   www.oshatools.com

---

## Heat Illness Prevention

Heat illness is a serious medical condition resulting from the body's inability to cope with increased heat load. Heat illness can be one or more medical conditions including: Heat Rash, Heat Cramps, Fainting, Heat Exhaustion, and Heatstroke. Heat Illness may be mild initially but can become severe or fatal if the body temperature continues to rise. Heat illness can also affect employees work performance and increase their risk of having accidents. Supervisors, foremen and employees should look continuously for signs and symptoms of Heat Illness in themselves and fellow workers. It is vital to immediately report any signs and symptoms of Heat Illness to a supervisor. There is a lot of variability in the recognition and reporting of heat illness symptoms.

---

A. Signs and symptoms of Heat Illness are:

1. Heat Rash (Prickly Heat) - a skin irritation caused by excessive sweating and clogged pores during hot, humid weather.

General Symptoms:

   a) Can cover large parts of the body.

   b) Looks like a red cluster of pimples or small blisters.

   c) Often on the neck, chest, groin, under the breasts, or in elbow creases.

   d) Uncomfortable, can disrupt sleep and work performance.

   e) Complicated by infections.

**Toolbox**

"Heat Illness Prevention"

**PUB210**   www.oshatools.com

2. Heat Cramps - Heat cramps affect people who sweat a lot during strenuous work activity. Sweating makes the body loose salts, fluids and minerals. If only the fluids are replaced and not the salts and minerals painful, muscles cramps may result.

General Symptom:

   a) Painful muscle spasms in the stomach, arms, legs, and other body parts may occur after work or at night.

3. Fainting - caused by a lack of adequate blood supply to the brain. Dehydration and lack of acclimatization to work in warm or hot environments can increase the susceptibility to fainting. Employees who stand for long periods or suddenly get up from a sitting or lying position when working in the heat may experience sudden dizziness and fainting. In both cases, victims normally recover consciousness rapidly after they faint.

General Symptoms:

a) Sudden dizziness, light-headedness.

b) Unconsciousness.

4. Heat Exhaustion - Heat exhaustion is the body's response to an excessive loss of the water and the salt contained in sweat. Older employees or those with high blood pressure are more susceptible to heat exhaustion. Cool skin temperature is not a valid indicator of a normal body temperature. Although the skin feels cool the internal body temperature may be dangerously high and a serious medical condition may exist.

General Symptoms:

a) Heavy sweating, painful muscle cramps, extreme weakness and/or fatigue.

b) Nausea, vomiting, dizziness, headache.

c) Body temperature normal or slightly high.

d) Fainting.

e) Pulse fast and weak.

f) Breathing fast and shallow.

g) Clammy, pale, cool, and/or moist skin.

**Toolbox**
"Heat Illness Prevention Plan (template)"
**PUB249**   www.oshatools.com

5. Heatstroke - Heatstroke is usually fatal unless emergency medical treatment is provided promptly. If the muscles twitch uncontrollably, keep the person from self-injury. Do not place any objects in the mouth. Monitor body temperature and continue cooling efforts until emergency medical treatment is provided to the victim.

General Symptoms:

a) No sweating, the body cannot release heat or cool down.

b) Mental confusion, delirium, convulsions, dizziness.

c) Hot and dry skin (e.g., red, bluish, or mottled).

d) Muscles may twitch uncontrollably.

e) Pulse can be rapid and weak.

f) Throbbing headache, shallow breathing, seizures and/or fits.

g) Unconsciousness and coma.

h) Body temperature may range from 102°-104° F or higher within 10-15 minutes.

B. Employers must protect employees from Heat Illness. All employees, foremen, and supervisors must be trained on the employer's heat illness prevention procedures. **3395(f)**

C. All employers, having employees exposed in outdoor places of employment, must have employer and site specific heat illness prevention plan. **3395**

---

**Toolbox**

"Protect Yourself from Heat Illness"

**PUB262**   www.oshatools.com

---

D. Heat illness prevention plan includes the following elements: **3395**

1. Access to drinking water. **3395(c)**

   a) Sufficient amounts of cool potable water shall be available at all times.

   b) Provide at least one quart per employee per hour for the entire shift.

c) Provide water at no cost to the workers.

d) Remind workers to drink water often and not to wait until they are thirsty to drink.

e) Place sufficient supplies of water as close to employees as possible given the worksite conditions and layout.

2. Shade requirements **3395(d)**

   a) When the outdoor temperature:

      (1) Does not exceed 85° F, provide shade or timely access to shade upon request.

      (2) Shade required to be present when the temperature exceeds 85° F. When the outdoor temperature in the work area exceeds 85° F, the employer shall have and maintain one or more areas with shade at all times while employees are present that are either open to the air or provided with ventilation or cooling. It is a good idea to set up the shade in advance, if at 5:00 p.m. the night before, the temperature is predicted to exceed 85° F. Or if you want to monitor the temperature during the work hours, perform hourly checks of the temperature at the worksite on the day of work and set up the shade immediately if the temperature exceeds 85° F.

      (3) Ever exceeds 90° F, at any time on the day of work, shade must be set up immediately.

   b) Place the shade structure as close as practicable to the areas where employees are working.

   c) Shade area should accommodate at least 25 percent of the employees on the shift at any time.

   d) Permit employees to access shade at all times.

   e) Encourage employees to take a cool-down rest in the shade, for a period of no less than 5 minutes at a time.

f) May provide alternative cooling measures that offer equivalent protection.

3. The employer shall use high-heat procedures when the temperature equals or exceeds 95° F. These procedures shall include the following to the extent practicable: **3395(e)**

   a) Ensuring that effective communication is maintained so that employees can contact a supervisor when necessary. An electronic device, such as a cell phone or text messaging device, may be used for this purpose only if reception in the area is reliable.

   b) Observing employees for alertness and signs or symptoms of heat illness.

   c) Reminding employees throughout the work shift to drink plenty of water.

   d) Close supervision of a new employee by a supervisor or designee for the first 14 days of the employee's employment by the employer (see exception in T8 CCR **3395(e)(4)**).

4. Written procedures that detail how the employer will:

   a) Provide access to water & shade.

   b) Monitor the weather.

   c) Institute high heat procedures and address lack of acclimatization.

   d) Train all employees and supervisors.

   e) Respond to heat illnesses without delay, provide first aid and emergency services.

   f) Provide clear and precise directions to the worksite.

5. Emergency Response procedure that shows how the employer will:

   a) Immediately respond to symptoms of possible heat illness.

   b) Contact emergency medical service providers.

c) Provide clear and precise directions to the worksite.

d) Ensure that emergency procedures are invoked when appropriate.

6. Training of all employees and supervisors on heat illness prevention before working outdoors in the heat. **3395(f)**

*Note: Training must include the importance of acclimatization, how it is developed, and how the employer's procedures address acclimatization.*

E. The employer must provide a suitable number of trained persons to render first aid as follows:

1. To give first aid for heat exhaustion, lay the person down flat in a cool environment, loosen his or her clothing, and give him or her plenty of water to drink.

2. To give first aid for heat stroke, immediately start aggressive cooling of the person and get him or her to a hospital right away.

F. Ways to prevent heat illness also include:

1. Monitoring weather forecast ahead of time and planning accordingly.

2. Timing the heaviest work load for during the coolest part of the workday.

3. Encouraging workers to drink water and to cool down.

4. Starting work shift early in the morning.

5. Providing training on heat stress including prevention, recognition, and first aid as a part of the employer's IIP Program. **3203**, **3400**, **3439**.

## Heavy Construction Equipment

Safety requirements for heavy construction equipment are as follows:

A. General repairs must not be made to powered equipment until workers are protected from movement of the equipment or its parts. **1595(a)**

B. Before repairs are made workers must comply with lock-out/block-out requirements if applicable. **3314**

C. Wherever mobile equipment operation encroaches on a public thoroughfare, a system of traffic controls must be used. **1598(a)**

D. Flaggers are required at all locations where barricades and warning signs can not control the moving traffic **1599(a)**. See exceptions in the "California Manual on Uniform Traffic Control Devices for Streets and Highways" (the Manual), published by CalTrans.

   Flaggers shall wear high visibility safety apparel and headwear manufactured in accordance to ANSI/ISEA standards **1599(d)**. Also, all employees (on foot), such as grade-checkers, surveyors and others exposed to the hazard of vehicular traffic, shall wear high visibility safety apparel in accordance with the requirements of **1598** and **1599**. **1590**

---

**Toolbox**

"Excavator and Backhoe Safety"

**PUB222**  www.oshatools.com

---

E. Job-site vehicles must be equipped with the following:

   1. Operable service, emergency, and parking brakes. **1591(c)**, **1597(a)**

   2. Two operable headlights and taillights for night operation. **1597(b)**

   3. Windshield wipers and defogging equipment as required. **1597(d)**

4. Seat belts if the vehicle has rollover protection structures. **1597(g)**

5. Fenders or mud flaps. **1591(f)**, **1597(i)**

6. Adequate seating if the vehicles are used to transport employees. **1597(f)**

---

**Toolbox**

"Injuries and Deaths from Skid-Steer Loaders"

**PUB226**   www.oshatools.com

---

F. Vehicles and systems must be checked for proper operation at the start of each shift. **1597(j)**

G. Rollover protection structures and seat belts must be installed for:

1. The following equipment having a brake horsepower rating above 20: **1596(a)(1)**

   a) Bulldozer

   b) Front-end loader

   c) Motor grader

   d) Scraper

   e) Tractor (except side boom pipe laying)

   f) Water wagon prime mover

---

**Toolbox**

"FMCSA Cargo Securement Rules"

**PUB243**   www.oshatools.com

---

2. The following equipment:

   a) Rollers and compactors (weighing more than 5,950 lbs.). **1596(a)(2)**

   *Exceptions: See* **1596(a)(2)(B)**

   b) Sheeps foot-type rollers and compactors. 1596(a)(2)(A)

   c) Crawler tractor. **3666**

---

H. Haulage and earth moving equipment safety requirements are as follows:

1. Every vehicle having a body capacity of 2.5 cu. yds. or more must be equipped with an automatic backup alarm that sounds immediately on backing. **1592(a)**

2. All other vehicles operating when rear vision is blocked must be equipped with an automatic backup alarm or its equivalent. **1592(b)**

3. All vehicles must be equipped with a manually operated warning device. **1592(c)**

4. Haulage vehicles in operation must be under operator control and must be kept in gear when descending grades. **1593(b)**

5. The brakes on a haulage vehicle must meet the criteria specified by the CSOs. **1591(c)**

6. The control devices on a haulage vehicle must be inspected at the beginning of each shift. **1593(d)**

7. Exposed scissor points on front-end loaders must be guarded. **1593(i)**

8. Engines must be stopped during refueling. **1594(a)**

9. Lights are required for night operation. **1591(g)**

10. Vehicles loaded by cranes, shovels, loaders, and similar devices must have an adequate cab or canopy for operator protection. **1591(e)**

11. Dust control is required when dust seriously limits visibility. **1590(b)**

12. In dusty operations, equipment operators shall use adequate respiratory protection. **1590(b)**

13. Loads on vehicles must be secured from displacement. **1593(f)**

---

**Toolbox**

"Analysis of Dozer Accidents"

**PUB217**   www.oshatools.com

---

I. Safety requirements for industrial trucks and tractors include:

1. Posting and enforcing by employers using industrial trucks or industrial tow tractors a set of operating rules including the appropriate rules listed in **GISO 3650(t)**. **3664(a)**

2. Providing operating instructions at the time of initial assignment and at least annually thereafter. **3664(b)**

3. Using the locking device where the dump body of a truck is raised for work. **1595(b)**

4. Performing repair work only when there is no possibility of sudden movements or operation of the equipment or its parts. Keeping all controls in a neutral position, with the engine(s) stopped and brakes set, unless work being performed requires otherwise. **1595(a)**

---

**Toolbox**

"Weather Impact on Large Truck Safety"

**PUB244** www.oshatools.com

---

## Hot Pipes and Hot Surfaces

Pipes or other exposed surfaces having an external surface temperature of 140° F (60° C) or higher and located within:

» 7 feet measured vertically from floor or working level, or

» 15 in. measured horizontally from stairways, ramps or fixed ladders shall be covered with a thermal insulating material or otherwise guarded against contact. **3308**

*Note: This order does not apply to operations where the nature of the work or the size of the parts makes guarding or insulating impracticable.* **3308**

---

## Housekeeping/Site Cleaning

Housekeeping is a term used to describe the cleaning of the work site and surrounding areas of construction project-related debris. The term also refers to the managing and storing of materials that are used on the project. Listed below are the general requirements for housekeeping to which all work sites are subject. It is important to remember that work sites subject to specific SOs may have additional housekeeping requirements with which to comply.

A. Work surfaces, passageways, and stairs shall be kept reasonably clear of scrap lumber and debris. **1513(a)**

B. Ground areas within 6 ft. of buildings under construction shall be kept reasonably free of irregularities. **1513(b)**

C. Storage areas and walkways on construction sites shall be kept reasonably free of dangerous depressions, obstructions, and debris. **1513(c)**

D. Piled or stacked material shall be placed in stable stacks to prevent it from falling, slipping, or collapsing. **1549(a)**

E. Material on balconies or in other similar elevated locations on the exteriors of buildings under construction shall be placed, secured or positively barricaded in order to prevent the material from falling. **1549(h)**

## Injury and Illness Prevention Program

An Injury and Illness Prevention (IIP) Program is required at all work sites. The program is considered effective if it satisfies the regulatory requirements of **3203** and helps the employer and the employee to identify and control the hazards specific to their work site. Following is a summary of the regulatory requirements.

A. The IIP Program must be in writing and must include the following elements: **1509(a)**, **3203(a)**

   1. The employer's assignment of responsibilities. **3203(a)(1)**

2. A system for ensuring employee compliance with safe work practices. **3203(a)(2)**

3. A system for two-way communication between employers and employees about safety issues. **3203(a)(3)**

4. Scheduled inspections and an evaluation system to identify hazards. **3203(a)(4)**

5. An accident investigation process. **3203(a)(5)**

6. Procedures for correcting unsafe and unhealthy conditions. **3203(a)(6)**

7. Safety and health training. **3203(a)(7)**

8. Recordkeeping. **3203(b)**

> **Toolbox**
> "IIPP for High Hazard Employers (template)"
> **PUB266**   www.oshatools.com

B. Other IIP Program requirements for construction are:

1. Employers must adopt and post a Code of Safe Practices at each job site. Plate A-3 in Appendix A of the CSOs illustrates a general format. **1509(b), (c)**

2. Periodic meetings of supervisors must be held to discuss the safety program and accidents that have occurred. **1509(d), 3203**

3. Supervisors must conduct tailgate or toolbox safety meetings at least every ten working days; however, weekly meetings are recommended. **1509(e)**

C. Required safety training for employees includes:

1. New workers shall be instructed in safe work practices, job hazards, and safety precautions and shall be required to read the Code of Safe Practices. **1510(a)**

2. The employer shall permit only qualified or experienced employees to operate equipment or machinery. **1510(b)**

3. Workers shall be instructed in the following:

   a) The recognition of job site-specific hazards.

   b) Procedures for protecting themselves.

   c) First aid procedures in the event of injury. **1510(c)**

---

**Toolbox**

"Collection of NIOSH Construction Research"

**PUB219**   www.oshatools.com

---

D. General safety requirements are as follows:

   1. No worker shall be required or permitted to work in an unsafe workplace. **1511(a)**

   2. Before starting work the employer shall survey the job site for hazards and use necessary safeguards to ensure that work is performed safely. **1511(b)**

E. Specific requirements are as follows:

   If an employer is subject to specific safety orders, the requirements of these SOs must be considered when developing the employer's IIP Program. These SOs may include specific procedures or processes as well as requirements for reporting, training, exposure limits, personal protection, and registration and certification.

F. Employees have numerous rights under the IIP Program, including the following: **3203(a)**

   1. The right to work in a safe and healthy workplace.

   2. The right to inform the employer of workplace hazards without fear of reprisal.

   3. The right to receive training that is readily understandable.

---

**Toolbox**

"Fatal Injuries to Workers"

**PUB224**   www.oshatools.com

---

G. To ensure the effectiveness of the IIP Program:

1. Supervisors should be qualified in safety procedures and held accountable.

2. The effectiveness of the safety program should be monitored.

## Ladders

Falls are the most common cause of worker injury associated with ladder use. Falls are mostly caused by the: (1) use of faulty ladders; (2) improper set-up of a ladder; or (3) incorrect use of ladders.

Except where either permanent or temporary stairways or suitable ramps or runways are provided, ladders shall be used to give safe access to all elevations. **1675(a)**

A. General requirements for ladders:

1. Portable ladders shall comply with T8 CCR **3276**. **1675(b)**

2. Design and construction of portable ladders shall comply with T8 CCR **3276(c).**

3. Fixed ladders shall comply with T8 CCR **3277**. **1675(c)**

4. Wood parts of fixed ladders shall meet the requirements of T8 CCR **3276(c)**. **3277(c)(5)**

5. Extension ladders shall comply with **3276(e)(16)**.

6. Portable metal ladders shall comply with **3276**.

7. Portable wood ladders shall comply with **3276**.

8. Portable reinforced plastic ladders shall comply with **3276(c)(e)**.

B. Portable ladders are generally designed for one-person use to meet the requirements of the person, the task, and the environment. When selecting a ladder for use, consider the ladder length, height, the working load, the duty rating, worker position, and how often the ladder is used. **3276(d)(1)(B)**

C. Double-cleat ladders are required for two-way traffic or when 25 or more employees are using a ladder. Double-cleat ladders shall not exceed 24 feet in length. **1629(c)**

D. Maximum lengths of portable ladders shall not exceed the following: **3276(e)(16)(D)**

| Ladder Type | Maximum Length (Feet) |
|---|---|
| Step ladder | 20 |
| Two-section extension ladder (wood) | 60 |
| Two-section extension ladder (metal) | 48 |
| Three-section extension ladder (metal) | 60 |
| Two-section extension ladder (reinforced plastic) | 72 |
| Painter's step ladder | 12 |
| Cleat ladder | 30 |
| Single ladder | 30 |

E. Minimum overlap in two section portable extension ladders shall not be less than the following: **3276(e)(16)(E)**

| Ladder Size (Feet) | Minimum Overlap (Inches) |
|---|---|
| Up to and including 32 | 36 |
| Over 32, up to and including 36 | 46 |
| Over 36, up to and including 48 | 58 |
| Over 48, up to and including 60. | 70 |

F. Portable ladders shall be used according to the following duty classifications: **3276(d)(2)**

| Duty Rating | Ladder Type | Working Load (Pounds) |
|---|---|---|
| Special Duty | IAA | 375 |
| Extra Heavy-Duty | IA | 300 |
| Heavy-Duty | I | 250 |
| Medium-Duty | II | 225 |
| Light-Duty | III | 200 |

G. All portable ladders used in outdoor advertising structures shall be at least Type I, Type IA or Type IAA as designed and constructed in accordance with T8 CCR **3276**. **3413(a)**

H. Job built Ladders

Job built ladders must meet the following requirements:

1. Job-built ladders must safely support the intended load. **1676(a)**

2. Cleats must be made from clear, straight-grained lumber and must be uniformly spaced 12 in. apart vertically. **1676(c)**

3. Cleats must be nailed at each end with three 10d nails or the equivalent. **1676(j)**

4. Cleats must be blocked or notched into the side rails. **1676(j)**

5. The width of single-cleat ladders shall be 15 in. to 20 in. **1676(f)**

6. Rails must be made from select Douglas Fir without knots (or the equivalent). **1676(b)**

7. Rail splicing is permitted only when there is no loss of strength to the rail. **1676(b)**

8. Single-cleat ladders must not exceed 30 ft. in length. **1676(e)**

9. Double-cleat ladders must not exceed 24 ft. in length. **1676(d)**

I. Portable Ladders

1. Inspection and maintenance requirements are below:

   a) Ladders shall be inspected by a qualified person for visible defects frequently and after any occurrence that could affect their safe use. **3276(e)(2)**

   b) Ladders shall be maintained in good condition at all times. **3276(e)(2)**

   c) Metal ladders shall not be exposed to acid or alkali materials that are capable of corroding the ladder and reducing the ladder's strength, unless recommended otherwise. **3276(e)(1)**

d) Remove ladders that have developed defects such as broken or missing steps, rungs, cleats, safety feet, side rails, or other defects from service, and tag or mark them with "Dangerous, Do Not Use". **3276(e)(3)**

e) All ladders shall be free of oil, grease, or slippery materials. Wood ladders shall not be painted with other than a transparent material. **3276(e)**

2. Prohibited uses of portable ladders are given below:

a) Ladders shall not be used as a brace, skid, guy or gin pole, gang-way, or for uses they were not intended, unless recommended by the manufacturer. **3276(e)(16)**

b) Do not place planks on the top cap. **3276(e)(16)(B)**

c) Step ladders shall not be used as single ladders or in the partially closed position. **3276(e)(16)(C)**

3. To safely use portable ladders employees must also follow the requirements noted below:

a) All portable ladders used for window washing shall be equipped with nonslip devices. Middle and top sections shall not be used as bottom sections unless equipped with nonslip bases. **3287(b)(2)**

b) Portable ladders shall not be overloaded when used. **3276(e)(6)**

c) The base of ladders shall be placed on a secure and level footing. Ladders shall not be placed on unstable bases. **3276(e)(7)**

d) Ladders shall not be used on ice, snow or slippery surfaces unless slippage is prevented. **3276(e)(7)**

e) The top of non-self-supporting ladder shall be placed with two rails supported equally, unless a single support attachment is provided and used. **3276(e)(8)**

f) Non-self-supporting ladders shall, where possible, be used so that the horizontal distance from the top support to the foot of the ladder is one-quarter of the working length of the ladder. **3276(e)(9)**

g) The ladder shall be so placed as to prevent slipping, or it shall be tied, blocked, held, or otherwise secured to prevent slipping. **3276(e)(9)**

h) Ladders shall not be used in a horizontal position as platforms, runways, or scaffolds unless designed for such use. **3276(e)(9)**

i) When two or more separate ladders are used to reach an elevated work area, the ladders shall be offset with a platform or landing between the ladders (see exceptions). **3276(e)(10)**

j) Extend ladder side rails to at least 3 ft. above the landing unless handholds are provided. **1629(c)(3), 3276(e)(11)**

k) Do not tie ladders together to provide longer sections unless the ladders are designed for such use and equipped with the necessary hardware fittings. **3276(e)(12)**

l) Extension ladders shall be erected so that the top section is above and resting on the bottom section with the rung locks engaged. **3276(e)(13)**

m) Do not place ladders in passageways, doorways, driveways, or any location where they may be displaced unless protected by barricades or guards. **3276(e)(14)**

n) Climb or work with the body near the middle of the step or rung and do not overreach from this position. To avoid overreaching, the employee shall descend and reposition the ladder. **3276(e)(15)(A)**

o) Employees shall be prohibited from carrying equipment or materials which prevent the safe use of ladders. **3276(e)(15)(B)**

p) Face the ladder while climbing and descending, and maintain contact with the ladder at three-points at all times. **3276(e)(15)(C)**

q) Do not stand and work on the top three rungs of a single or extension ladders. **3276(e)(15)(D)**

r) Employees shall not stand on the topcap or the step below the topcap of a stepladder. **3276(e)(15)(E)**

s) Do not use the X-bracing on the rear section of a stepladder for climbing unless the ladder is so designed and provided with steps for climbing on both front and rear sections. **3276(e)(15)(F)**

t) Ladders shall not be moved or extended while occupied, unless designed and recommended by the manufacturer. **3276(e)(15)(G)**

u) Portable rung ladders with reinforced rails shall be used only with the metal reinforcement on the under side. **3276(e)(17)**

v) Non-conductive ladders shall be used in locations where the ladder or user may contact unprotected energized electrical conductors or equipment. Conductive ladders shall be legibly marked with signs reading "CAUTION-- DO NOT USE AROUND ELECTRICAL EQUIPMENT," or equivalent. **3276(e)(18)**

w) The area around the top and bottom of a ladder shall be kept clear. **3276(e)(19)**

**Toolbox**
"Contacting Power Lines with Metal Ladders"
**PUB220** www.oshatools.com

J. Fixed Ladders

To safely use fixed ladders, employees must also follow the requirements noted below:

1. Do not carry equipment or materials which prevent the safe use of ladders. **3278(a)**

2. Fixed ladders shall be inspected before use. Any ladder determined to be unsafe shall not be used. **1511(b)**

3. Face the ladder when ascending and descending. **3278(a)**

4. Always using both hands when climbing up or down the ladder. **3278(a)**

5. Do not use single-rail ladders. **3278(a)**

K. The following are training requirements for employees using portable ladders: **3276(f)**

1. Employees shall be trained in the safe use of ladders before using them.

2. Supervisors of employees who routinely use ladders shall also be trained in ladder safety training.

3. The training may be provided as part of the employer's IIP Program (**T8 CCR 3203**).

4. The training shall address the following topics, unless the employer demonstrates that a topic is not applicable to the workplace:

a) Importance of using ladders safely, including injuries due to falls from ladders. **3276(f)(1)**

b) Selection of ladders, including types, proper length, maximum working loads, and electrical hazards. **3276(f)(2)**

c) Maintenance, inspection, and removal of damaged ladders from service. **3276(f)(3)**

d) Erecting ladders including footing support, top support, securing, and angle of inclination. **3276(f)(4)**

e) Climbing and working on ladders including user's position and points of contact with the ladder. **3276(f)(5)**

f) Causes of falls, including haste, sudden movement, lack of attention, footwear, and user's physical condition. **3276(f)(6)**

g) Prohibited uses including climbing on cross bracing, uses other than designed, exceeding maximum lengths, and not meeting minimum overlap requirements. **3276(f)(7)**

L. It is a good idea to make sure that the stepladder is properly set up and that the spreader is in locked position before use.

## Laser Equipment

The primary hazard of using laser equipment is injury to the eyes. Following are selected regulatory requirements:

A. Only qualified persons may operate laser equipment. **1801(a)**

B. Equipment must be turned off or shielded when unattended and not in use. **1801(e)**

C. Laser beams must never be pointed or directed at persons. **1801(g)**

D. Lasers must have a label indicating their maximum output. **1801(i)**

E. Employees who have a potential exposure to direct or reflect laser light greater than 5 milliwatts shall be provided with anti-laser eye protection as specified in Section **3382(e)**. **1801(c)**

F. Warning signs and labels (in accordance with ANSI) must be posted in areas where lasers are used. **1801(d)**

# Lead

Occupational exposures to lead can occur in construction activities, such as plumbing system retrofits; the spraying, removal, or heating of paint that contains lead; and the welding, cutting, and grinding of lead-containing construction materials.

Occupational lead exposures can affect workers as well as family members and friends who come into contact with the "take -home" lead on the worker's clothing, hair, hands, etc. The toxic effects of lead on the human body have been well documented and include damage to the kidneys, brain, and reproductive organs that in turn causes the loss of kidney function, sterility, decreased fertility, and birth defects.

Because of the serious and, in many cases, life-threatening health effects of lead, the employer must be thoroughly knowledgeable about the regulations to protect people from lead exposure before their employees engage in any work exposing them to lead. **1532.1**

**Toolbox**

"Lead in Construction"

**PUB261**   www.oshatools.com

A. Cal/OSHA enforces the "Lead in Construction Safety Orders" that makes employers responsible for the following: **1532.1.**

1. For each job site the lead hazard must be assessed. **1532.1(d)(1)**

2. Where lead is present the following is required:

   a) Lead dust must be controlled by HEPA vacuuming, wet cleanup, or other effective methods. **1532.1(h)**

b) The employer shall assure that food, beverage, and tobacco products are not present or used in areas where employees are exposed to lead above the PEL. The employer shall provide hygiene facilities for changing, showering, eating, and hand washing. **1532.1(i)**

c) Workers shall receive appropriate training. **1532.1(l)**

d) The employer shall implement a written compliance program to ensure control of hazardous lead exposures. **1532.1(e)**

e) The employer shall provide the worker with and require the use of appropriate personal protective equipment. **1532.1(f),(g)**

f) The employer shall assure that all protective clothing is removed at the completion of a work shift only in change areas provided for that purpose. **1532.1(g)**

B. The permissible exposure limits (PELs) for airborne lead are 0.05 milligrams per cubic meter of air (mg/m3) and an action level of 0.03 mg/m3, both as an 8 hour time-weighted-average (TWA). **1532.1(b),(c)**

C. Trigger tasks are certain highly hazardous tasks that carry the presumption of airborne exposure above the PEL. They require special protective measures until it is determined that worker airborne exposures to lead are below levels specified in **1532.1**.

Following are the three levels of trigger tasks involving lead-containing materials and associated respirator requirements: **1532.1(d)(2)**

1. Level 1 trigger tasks: spray painting, manual demolition, manual scraping or sanding, using a heat gun, and power tool cleaning with dust collection system.

   » Minimum respirator requirement: a half-mask respirator with N100, R100, or P100 filters.

2. Level 2 trigger tasks: using lead containing mortar; burning lead; rivet busting; cleaning power tools without a dust collection system; using dry, expendable abrasives for clean-up procedures; moving or removing an abrasive blasting enclosure.

   » Minimum respirator requirement: a full-face mask respirator with N100, R100, or P100 filters; a supplied-air hood  or helmet; or a loose-fitting hood or helmet with a powered air purifying respirator with N100, R100, or P100 filters.

3. Level 3 trigger tasks: abrasive blasting, welding, cutting, or torch burning on structures.

   » Minimum respirator requirement: a half mask, supplied-air respirator operated in a positive pressure mode.

D. Protective requirements for all trigger tasks and any other task that may cause a lead exposure above the PEL include the following:

1. Respirators, protective equipment, and protective clothing. **1532.1(f), (g)**

2. Clothing change areas and a shower. **1532.1(i)**

3. Initial blood tests for lead and zinc protoporphyrin. **1532.1(j)**

4. Basic lead hazard, respirator, and safety training. **1532.1(l)**

5. The establishment of a regulated area and warning signs as shown below: **1532.1(i), (m)**

# WARNING

# LEAD WORK AREA

# — POISON —

# NO SMOKING OR EATING

*Note: The above protective requirements must be enforced until worker airborne exposures are shown to be below levels specified in 1532.1.*

E. Blood lead monitoring is especially important to evaluating work and hygiene practices that may result in lead ingestion. Employees whose blood lead levels exceed specified limits must be removed from the work with exposure to lead at or above the action level. These workers must be provided with normal earnings, seniority, and other employee rights and benefits for 18 months or until the job from which they were removed is discontinued, whichever occurs first. Mandatory medical removal of an employee due to lead (or other regulated chemicals) must be recorded on the Log 300 with a check in the "poisoning" column. **1532.1(k)(2)**, **14300.9**

F. Feasible engineering and work practice controls must be implemented to maintain employee exposures to lead below the PELs.

G. A written compliance program that details how lead exposures will be controlled is required. **1532.1(e)**

H. On jobs at residential and public-access buildings, workers whose exposures to lead measure above the PELs, and their supervisors, must receive state-approved training and certification by the California Department of Health Services.

I. Records of air monitoring, blood lead testing, and medical removal must be maintained. **1532.1(n)**

J. Employers who conduct lead work listed in **1532.1(d)(2)** must notify the Division, in writing, at least 24 hours before the start of work. **1532.1(p)**

# Lighting

A. Proper illumination is important in all construction activities. Construction areas, ramps, corridors, offices, shops and storage areas, etc., shall be lighted to not less than the minimum illumination intensities in the following table while work is in progress: **1523(a)**

Minimum Illumination Intensities In Foot-Candles

| Foot-Candles | Area or Operation |
|---|---|
| 3 | General construction area lighting low activity. |
| 5 | Outdoor active construction areas, concrete placement, excavation and waste areas, access ways, active storage areas, loading platforms, refueling, and field maintenance areas. |
| 5.. | Indoors: warehouses, corridors, hallways, stairways, and exit-ways. |
| 10... | General construction plant and shops (e.g., batch plants, screening plants, mechanical and electrical equipment rooms, carpenter shops, rigging lofts and active storerooms, barracks or living quarters, locker or dressing rooms, mess halls and indoor toilets and workrooms). |
| 10.... | Nighttime highway construction work. |
| 30..... | First-aid stations, infirmaries, and offices. |

B. Nighttime highway construction work lighting shall be provided within the work zone to illuminate the task(s) in a manner that will minimize glare to work crews and not interfere with the vision of oncoming motorists. **1523(b)**

# Lock-out/Block-out Procedures

Every year many employees are injured or lose their lives when the equipment they are repairing or maintaining is turned on by a coworker or when potential energy is released while the employee is in harm's way of the equipment.

A. When equipment needs to be de-energized during cleaning, servicing, or adjusting operations the following applies: **GISO 3314**

   1. Machinery or equipment capable of movement shall be stopped, and the power source shall be de-energized or disengaged. **3314(c)**

   2. Moveable parts shall be mechanically blocked or locked out. **3314(c)**

   3. Equipment that has lockable controls or that is readily adaptable to lockable controls shall be locked out or positively sealed in the off position. **3314(d)**

   4. Accident prevention signs or tags shall be placed on the controls of equipment, machines, and prime movers during repair work. **3314(c)**

   5. An energy control procedure shall be developed and used by the employer. **3314**

B. If the equipment must move during repair or maintenance, the employer shall provide and require the use of extension tools or other means to protect employees from injury due to the movement. Employees shall be trained on the safe use and maintenance of such tools or means. **3314(c)(1)**

C. For heavy construction equipment repair, **1595(a)** requires that repairs must not be made until workers are protected from movement of the equipment or its parts.

D. An authorized person shall be responsible for the following before working on de-energized electrical equipment or systems unless the equipment is physically removed from the wiring system: ESO **2320.4**

   1. Notifying all involved personnel. **2320.4(a)(1)**

2. Locking the disconnecting means in the "open" position with the use of lockable devices, such as padlocks, combination locks or disconnecting of the conductor(s) or other positive methods or procedures which will effectively prevent unexpected or inadvertent energizing of a designated circuit, equipment or appliance. **2320.4(a)(2)**

   *Exception: Locking is not required under the following conditions:*

   a) Where tagging procedures are used as specified in **2320.4(a)(3)**, and

   b) Where the disconnecting means is accessible only to personnel instructed in these tagging procedures.

3. Tagging the disconnecting means with suitable accident prevention tags conforming to the provisions of **2320.6** and **3314(e)**. **2320.4(a)(3)**

4. Effectively blocking the operation or dissipating the energy of all stored energy devices which present a hazard, such as capacitors or pneumatic, spring-loaded and like mechanisms. **2320.4(a)(4)**

---

**Toolbox**

"Lockout / Blockout"

**PUB260**   www.oshatools.com

---

# Machine Guarding

Machine guarding is required on all moving machine parts when the operation of a machine or accidental contact with the parts could injure the operator or other workers. The following are some of the major moving machine parts that must be guarded:

» Gears, sprockets, and chain drives. **4075(a)**

» Belt and pulley drives. **4070(a)**

» Belt conveyor head and tail pulleys. **3999(b)**

---

- » Screw conveyors. **3999(a)**
- » Exposed shafts and shaft ends. **4050(a)**, **4051(a)**
- » Collars and couplings. **4050(a)**
- » Hazardous revolving or reciprocating parts. **4002(a)**

> **Toolbox**
>
> "Machine Guarding"
>
> **PUB238**   www.oshatools.com

## Multi-employer Work Sites

Multi-employer work sites are work locations where more than one employer and his or her employees work, usually but not necessarily at the same time. Most construction sites are multi-employer work sites, and therefore more than one employer is responsible for safety at these work sites. Each employer is required to notify the other employers of hazards and to guard against exposing their own employees as well as all other employees on the site.

The four categories of employers who may be cited by Cal/OSHA for employee exposures to violative conditions are identified in **336.10, 336.11**:

A. Exposing Employer is an employer whose employees were exposed to the violative condition at the work site regardless of whether that employer created the violative condition.

B. Creating Employer is an employer who actually created the violative condition.

C. Controlling Employer is an employer who is responsible, by contract or through actual practice, for safety and health conditions at the work site and who has the authority to correct the violation.

D. Correcting Employer is an employer who has the responsibility to correct the violative condition.

# Personal Protective Equipment

When a hazard cannot be eliminated or controlled by engineering or administrative controls as required by Cal/OSHA regulations, workers must be protected by personal protective equipment (PPE) as follows:

A. Eye and face protection is required when there is an inherent risk of eye injury from flying particles, injurious chemicals, or harmful light rays. **3382**

B. Foot protection is required for workers who are exposed to foot injury from hot, corrosive, or injurious substances; from falling objects; or from crushing or penetrating actions. Foot protection is also required for employees who work in abnormally wet locations. **3385**

C. Hand protection is required for workers who are exposed to cuts, burns, electrical current, or harmful physical or chemical agents. **1520, 2320.2(a)**

D. Body protection is required for workers who are exposed to injurious materials. These workers must wear appropriate body protection and clothing appropriate for their work. **1522(a)**

   1. Loose clothing, such as sleeves, ties, and cuffs, may not be worn around machinery in which it could become entangled. **1522(b)**

   2. Workers must not wear clothing saturated with flammable liquids or corrosive or oxidizing agents. **1522(c)**

---

**Toolbox**

"Protecting Yourself – Noise in Construction"

**PUB213**   www.oshatools.com

---

E. Hearing protection (HP) is required because the noise levels of many construction operations frequently exceed 90 dBA. When employees are subjected to sound levels listed in Table 3 (**5096(b)**), feasible administrative or engineering controls must be used. If these controls fail to reduce sound levels to an acceptable range, workers must wear hearing protection and be trained to properly use the HP devices.

## Table 3
### Allowable Exposure Levels to Sound

| Sound level (dBA) | Time per day (hours) |
|:---:|:---:|
| 90 | 8 |
| 95 | 4 |
| 100 | 2 |
| 105 | 1 |
| 110 | $^1/_2$ |

F. Head protection is required for employees who are exposed to flying or falling objects or to electric shocks and burns. These employees must wear approved head protection. Hair must be confined if there is a risk of injury from entangling it in moving parts, combustibles, or toxic contaminants. **3381(a)**

*Note: Everyone at a construction site should wear hard hats with bills in the forward position.*

G. Respiratory protection is required when engineering or administrative controls are not feasible for limiting harmful exposure to airborne contaminants. In these circumstances exposed employees must wear respirators approved by the National Institute for Occupational Safety and Health (NIOSH). **5144(a)**

For all respirator use a written respiratory protection program must be in place, covering employee training, respirator selection, medical evaluation, fit testing, use, cleaning, sanitizing, inspection, and maintenance. **5144(a),(c)**

H. Personal flotation devices are required to be worn when working over or near water. **1602**

I. Some of the SOs require specialized personal protective equipment not mentioned here. Employers and employees should refer to the specific SOs applicable to the type of work they perform to determine additional PPE requirements (for example the Electrical Safety Orders **2299 – 2874**).

J. Work on exposed energized parts of equipment or systems is allowed when suitable personal protective equipment and safeguards (i.e., approved insulated gloves or insulated tools) are provided and used, and other conditions as listed in **2320.2(a)** are met. **2320.2**

## Pile Driving

Regulations concerning pile driving are as follows:

A. A supervised danger zone must be established around the operating hammer if employees are cutting, chipping, or welding. **1600(a)**

B. A blocking device or other effective means capable of safely supporting the weight of the hammer shall be provided to secure the hammer in the leads and shall be used at all times while when any employees are is working under the hammer. **1600(b)**

C. All pressurized lines and hoses must be secured by 1/4 in. alloy steel chain (3250 pound rated capacity) or wire rope of equivalent strength. **1600(c)(1)**

D. When used, work platforms must meet the specific requirements of **1600(d)**.

E. Leads shall be provided with a continuous ladder or horizontal bracing that is uniformly spaced at intervals no greater than 18 inches, and the leads shall be equipped with adequate anchorages for use with a personal fall protection system. The operator of the equipment will apply all brakes and necessary safety switches to prevent uncontrolled motion of the equipment before an employee may access the leads. **1600(f)**

F. Fall protection must be provided when workers are exposed to unguarded platforms or walkways exceeding 7 1/2 ft. in height. **1670(a)**

G. Walkways that are at least 20 in. wide must be provided for access to all work areas. **1600(h)**

H. Employees shall not ride the hammer, crane load block or overhaul ball. Sheet piling shall be firmly stabilized before workers are permitted to work on them. **1600(g)**

I. Where a drop hammer is used for driving piling, other than sheet piling, a driving head or bonnet shall be provided to bell the head of the pile and hold it true in the leads. **1600(h)(3)**

The pile hammer, clamp, power unit and supply hoses shall be inspected in accordance with their manufacturer's recommendations. **1600(i)**

J. Adequate and accessible flotation gear (ring buoys, a life saving boat) must be provided to protect workers who are exposed to a drowning hazard. **1600(j),(k)**

K. The engine or winch operator shall receive signals only from a designated signaler. **1600(l)**

*Exception: When an employee is aloft in the leads, the hammer shall not be moved except on the signal of the employee aloft.*

L. A hammer stop block is required. **1600(o)**

M. Two steam (or compressed air) shutoff valves are required; one must be a quick-acting valve within reach of the hammer operator. **1600(c)(2)**

N. Rigs must be stabilized with guys or outriggers when needed. Hammers shall be lowered to the bottom of the leads while the pile driver is being moved (traveling). **1600(p)**

O. Piles shall be unloaded and stored in a controlled manner. **1601**

P. The rated capacity of the hammer's suspension shall not be exceeded. The manufacturer's recommendations for extracting piling shall be observed at all times. **1600.1**

## Pressurized Worksites

Pressurized worksites (also known as compressed-air worksites) are sites where employees perform duties in a pressurized environment, such as a caisson. Employees working on pressurized worksites may be exposed to some specific health and safety hazards due to compression and decompression. These hazards are similar to hazards found in diving operations, pressurized tunneling operations, and in confined spaces. Employees may develop decompression sickness (bends) from exposure to decompression. The symptoms of decompression sickness include headache, unusual tiredness, rash, pain in one or more joints, tingling in the arms or legs, muscular weakness or paralysis, breathing difficulties, shock, unconsciousness or death.

Also, in a pressurized work environment structural failures or blowouts may occur. This may lead to the work area becoming inundated with mud and water causing drowning and asphyxia.

A. Cal/OSHA must receive written notification at least seven days before the work is started. **6075**

B. Regulatory requirements for pressurized (hyperbaric) work environments include:

1. Following the guidelines for proper compression of employees as per **6080(a)**.

2. Not subjecting employees to pressure exceeding 50 pounds per square inch. **6080(b)**

3. Not allowing employees working in compressed air to pass from the working chamber to atmospheric pressure until after decompression, in accordance with **6085. 6090**

*Exception: The requirements above do not apply in an emergency. 6080*

4. Controlling decompression of employees as discussed in **6085**.

5. Decompression of employees in accordance with the specified decompression tables of the U.S. Navy Diving Manual, Volume 2, Chapter 9, 2008. **6085**

*Exception: Requirements #4 and #5 do not apply in an emergency provided that employees are decompressed in accordance with decompression tables and procedures recommended by the supervising physician.*

6. Temperature, illumination, sanitation, and ventilation as per **6100**. Ventilation in the locks and chambers, with the exception of the medical chamber, shall be such that the air quality meets the requirement of Section **5144(i)**. Ventilating air shall be not less than 30 cubic feet per minute per person. **6100**

7. Providing forced ventilation during decompression to ensure a source of fresh air. **6100(f)**

8. Taking one, or both, of the following steps when an oxygen breathing gas system is used during decompression, to ensure that the concentration of oxygen inside the chamber or lock does not exceed twenty five percent (25%) by volume: **6100(i)**

   a) The oxygen breathing gas system shall capture the oxygen that is not consumed by the user and directly exhaust it to a well ventilated area outside of the lock or chamber.

   b) An oxygen meter shall be used to continuously monitor the oxygen concentration inside the chamber or lock.

9. The employer retaining a supervising physician who shall be available at all times while pressurized work is in progress in order to provide medical supervision of employees employed in compressed air work. **6120.**

10. Following fire prevention and oxygen safety requirements as specified in **6115.**

C. Employees who are exposed to or control the exposure of others to hyperbaric conditions shall be trained in hyperbaric related physics and physiology, recognition of pressure related injuries, and how to avoid discomfort during compression. **6075**

## Qualified Person

A qualified person is a person designated by the employer; and who by reason of training, experience, or instruction has demonstrated the ability to perform safely all assigned duties; and, when required, is properly licensed in accordance with federal, state, or local laws and regulations. **1504** The CSOs refer to a qualified person in several of the regulations.

## Ramps and Runways

Regulations concerning ramps and runways are as follows:

A. General requirements:

1. Ramps must be properly designed to provide a safe means of access for foot or vehicle traffic. **1623**, **1624**, **1625**

2. Open sides of ramps that are 7 1/2 ft. or more above ground must have standard guardrails. **1621(a)**

B. Foot ramps:

1. Foot ramps must be at least 20 in. wide and must be secured and supported to avoid deflection or springing action. **1624(a)**

2. If the ramp slope exceeds 2 ft. of rise for every 10 ft. of run, cleats must be 8 in. or more in length and must be placed not more than 16 in. apart. **1624**

C. Wheelbarrow ramps and runways:

1. Wheelbarrow ramps and runways must be firmly secured against displacement. **1624(c)**

2. Ramps more than 3 ft. high must be 30 in. wide, and planks must be firmly cleated together. **1623**

3. Falsework design loads must be increased by 10 psf for worker-propelled carts. **1717(a)**

## Roofing Operations

Working conditions at roofing projects are often difficult and continuously expose workers to serious hazards. In California one of the most common causes of work-related deaths is falls from roofs. Injuries common to the roofing industry include (1) broken bones because of falls; (2) back injuries because of awkward postures and heavy lifting; and (3) burns from contact with hot roofing asphalt and associated equipment.

Roofing operations are classified as either single-unit or multi-unit. Examples of single-unit (monolithic) roofing are built-up roofing, flat-seam metal roofing, and vinyl roofing. Examples of multi-unit roofing are asphalt shingles, cement, clay and slate tile, standing seam metal panels, shingle metal roofing, and wood shingles.

Employees shall be protected from falls from roofs. The following regulations aim to minimize or eliminate the hazards associated with the roofing industry:

A. Specific fall protection methods are used for: **1730**

» Different roof heights and slope conditions.

» Different types of roofing operations including custom-built homes.

» Re-roofing operations.

» Roofing replacements or additions on existing residential dwelling units.

» Roofing operations (including new production-type residential construction) with slopes less than 3:12.

1. For single-unit roofs with slopes of 0:12 through 4:12 and more than 20 ft in height: **1730(b)**

   a) Warning lines and headers. **1730(b)**

   b) Personal fall protection systems per **1724(f)**.

   c) Catch platforms with guardrails. **1724(c)**

   d) Scaffold platforms. **1724(d)**

   e) Eave barriers. **1724(e)**

   f) Parapets that are 24 in. or higher. **1730(b)**

   g) Standard railings and toeboards. **Article 16**

Exceptions: **1730(b)**

» Whenever any equipment is pulled by an operator who walks backwards, one or a combination of the above methods shall be applied regardless of height.

» At those job sites where any equipment is pulled by an operator who walks backwards or an operator rides motorized equipment the parapet must be 36 in. or more in height at those roof edges which are perpendicular (or nearly so) to the direction in which the equipment is moving.

**Toolbox**

"Worker Deaths by Falls"

**PUB235**   www.oshatools.com

2. For single-unit roofs with slopes exceeding 4:12 and more than 20 ft in height: **1730(c)**

   a) Parapets that are 24 in. or higher. **1730(c)**

   b) Personal fall protection systems per **1724(f)**.

   c) Catch platforms. **1724(c)**

   d) Scaffold platforms. **1724(d)**

   e) Eave barriers. **1724(e)**

   f) Standard railings and toeboards. **Article 16**

Exception:

*Provisions in **1730(c)** do not apply at job sites where the motorized equipment on which the operator rides:*

   » Has been designed for use on roofs having slopes greater than 4:12, and

   » Is used where a parapet is:

      i.  At least 36 in. high at roof edges, and

      ii. Perpendicular to the direction in which the equipment is moving.

3. For single-unit roofs with slopes exceeding 4:12, no equipment that is pulled by an operator walking backwards shall be used.

4. For multi-unit roofs with slopes 0:12 through 5:12 and more than 20 ft in height, employees shall be protected from falls by the use of one of the following: **1730(c)**

   a) A roof jack system as provided in Section **1724(a)**.

   b) A minimum of 24 inch high parapet.

   c) Other methods affording equivalent protection.

5. For multi-unit roofs with slopes exceeding 5:12 and more than 20 ft in height, employees shall be protected from falls by the use of one or a combination of the following: **1730(f)**

   a) Parapets that are at least 24 in. high.

   b) Personal fall protection systems per **1724(f)**.

   c) Catch platforms. **1724(c)**

   d) Scaffold platforms. **1724(d)**

   e) Eave barriers. **1724(e)**

   f) Roof jack systems. (Safety lines are required when using roof jack systems on roofs steeper than 7:12.) **1724(a)**

B. New production-type residential construction with roof slopes of 3:12 or greater have specific fall protection requirements. **1731**

   1. For New Production-Type Residential Construction with slopes 3:12 through 7:12 and the eave height exceeds 15 feet above the grade or level below, employees shall be protected from falling when on a roof surface by use of one or any combination of the following methods:

      a) Personal Fall Protection. **1670**

      b) Catch Platforms. **1724(c)**

      c) Scaffold Platforms. **1724(d)**

      d) Eave Barriers. **1724(e)**

      e) Standard Railings and Toeboards. **Article 16**

      f) Roof Jack Systems. **1724(a)**

   2. For New Production-Type Residential Construction with slopes greater than 7:12 regardless of height, employees shall be protected from falling by methods prescribed in the above subsections a, b, c, and e. **1731(c)**

C. Roofing operations require documented employee training. For New Production-Type Residential Construction, training shall include the followings in addition to those required by **1509** and **3203**:

1. Work on or near gable ends.

2. Slipping hazards.

3. Roof holes and openings.

4. Skylights.

5. Work on ladders and scaffolds.

6. Access to the roof.

7. Placement and location of materials on the roof.

8. Impalement hazards.

9. Care and use of fall protection systems.

D. Hot operations are subject to the following regulations:

1. Workers must not carry buckets containing hot material up ladders. **1725(a)**

2. An attendant must be stationed within 100 ft. of any kettle not equipped with a thermostat. **1725(d)**

3. Liquefied petroleum gas cylinders must not be located where the burner will increase the temperature of the cylinder. **1725(g)**

4. A Class BC fire extinguisher shall be kept near each kettle in use as shown below:

    a) For a kettle with a capacity of less than 150 gal. = 8:BC

    b) For a kettle with a capacity of 150 gal. to 350 gal. = 16:BC

    c) For a kettle with a capacity of more than 350 gal. = 20:BC. **1726(d)**

5. The fuel tanks of compressed-air-fueled kettles must be equipped with a relief valve set for a pressure not to exceed 60 psi. **1726(c)**

6. Coal tar pitch operations are subject to the following requirements:

   a) Workers must use skin protection. **1728(a)**

   b) Washing or cleansing facilities must be available. **1728(c)**

   c) Workers must use respirators and eye protection in confined spaces that are not adequately ventilated. **1728(b), 5158**

7. Hot pitch and asphalt buckets have the following maximum capacities:

   a) Carry buckets = 6 gal.

   b) Mop buckets = 9 1/2 gal. **1729(a)(2),(4)**

E. Personal fall protection for roofing operations is regulated as follows: **1724(f)**

1. Personal fall arrest systems, personal fall restraint systems, and positioning devices must be installed and used in accordance with Article 24 in the GISO. **1724(f)**

2. Safety lines must be securely attached to substantial anchorages on the roof. **1724(f)**

3. Roof openings must be railed or covered. Temporary railing and toeboards shall meet the requirements of Sections **1620** and **1621**. The railing shall be provided on all exposed sides, except at entrances to stairways. **1632(b)(2)**

4. The cover must be securely fastened and able to withstand 2 times the expected load or a minimum of 400 pounds. Covers must bear a sign stating - OPENING-DO NOT REMOVE. **1632(b)(3)**

5. An employee approaching within 6 feet of any finished skylight or skylight opening must be protected from falling through the skylight or opening as specified in **3212(e)**.

## Scaffolds

Work activities associated with scaffolds are subject to many hazards, however falls are by far the number-one cause of injury or death among construction workers. The following requirements regulate the design, erection, use and dismantling of scaffolds:

A. General requirements:

1. Scaffolds must be provided for work that cannot be done safely by employees standing on ladders or on solid construction that is at least 20 in. wide.

   *Exception: A 12-inch wide plank on members that are on 24 inch (or closer) centers is permitted.* **1637(a)**

2. The design and construction of scaffolds must conform to applicable standards and requirements. **1637**, ANSI A10.8-1988, ANSI/ASSE A10.8-2001 Standards are based on stress grade lumber. Metal or aluminum may be substituted if the structural integrity of the scaffold is maintained. **1637(b)**

3. Manufactured scaffolds shall be used in accordance with the manufacturer's recommendations. **1637(b)(4)**

   *Exception: Where specific requirements that address riding on a rolling scaffold in Section* **1646(i) and (j)** *may conflict with the manufacturer's recommendations, the provisions in Section* **1646(i) and (j)** *take precedence.*

   **Toolbox**
   "Working Safely With Supported Scaffolds"
   **PUB258**   www.oshatools.com

4. Each scaffold must be designed to support its own weight and 4 times the maximum load. Maximum working loads are as follows: **1637(b)**

   a) Light-duty scaffolds: 25 psf of work platform.

   b) Medium-duty scaffolds: 50 psf of work platform.

c) Heavy-duty scaffolds: 75 psf of work platform.

d) Special-duty scaffolds: exceeding 75 psf as determined by a qualified person or a California registered Civil Engineer with scaffold design experience.

e) Engineered scaffolds: as determined by a California registered Civil Engineer with scaffold design experience.

5. The erecting and dismantling of scaffolds are regulated as follows:

   a) Scaffold erection and dismantlement must be supervised by a qualified person. **1637(k)(1)**

   b) Scaffolds must be erected and dismantled according to design standards, engineered specifications, or manufacturer's instructions. **3328, 1637(k)**

   c) A DOSH permit is required for erecting and dismantling scaffolds that exceed three stories or 36 ft. in height. **341(d)(5)(B)**

6. Scaffold access: Ladders, horizontal members, and stairways must provide safe and unobstructed access to all platforms.

   The equipment must be located so that its use will not disturb the stability of the scaffold: **1637(n)**

   a) Ladders may be used if the following applies:

      (1) Portable ladders shall comply with T8 CCR 3276. **1675(b)**

      (2) Fixed ladders shall comply with T8 CCR 3277. **1675(c)**

      (3) Ladders must be securely attached to scaffolds. **1637(n)**

      (4) Ladders must extend 3 ft. above the platform, or handholds must be provided. **3276(e)(11)**

   b) Manufactured hook-on and attachable ladders shall be securely attached to the scaffold and: **1637(n)**

(1) Shall be specifically designed for the type of scaffold used;

(2) Shall have a minimum rung length of 11-1/2 in. (29 cm); and

(3) Shall have uniform spaced rungs with a maximum spacing between rungs of 16- 3/4 in.

c) Horizontal members built into the end frame of a scaffold may be used to access platforms if: **1637(n)**

(1) The horizontal members are parallel and level.

(2) The horizontal members make a continuous ladder, bottom to top, with the ladder sides of the frames in a vertical line.

(3) The horizontal members provide sufficient clearance for a good handhold and foot space. **1637(n)**, **1644(a)**

d) Stairways must conform to the following: **1637(n)(2)**

(1) Permanent stairways for scaffolds must comply with GISO requirements (for example **3214**, **3622**).

(2) Prefabricated scaffold steps or stairs must comply with:

- ANSI 10.8-1988 or ANSI/ASSE 10.8-2001 if manufactured on or before May 28, 2005

- ANSI/ASSE 10.8-2001 if manufactured after May 28, 2005

**Toolbox**

"Scaffold Use in the Construction Industry"

**PUB214**   www.oshatools.com

7. Scaffolds must be secured as follows:

a) Scaffolds must be tied off with a double- looped No. 12 iron wire or a single- looped No. 10 iron wire or the equivalent. A compression member should prevent scaffold movement toward the structure. **1640**, **1641**, **1644**

b) Light duty wooden pole scaffolds must be tied off every 20-ft. horizontally and vertically. **1640(b)**

c) Heavy-trade wooden pole scaffolds must be tied off every 15-ft. horizontally and vertically. **1641(f)**

d) Metal scaffolds must be tied off as specified in **1644(a)(5)**.

8. Scaffold platforms must conform to the following:

a) Platforms must be capable of supporting the intended load. **1644(a)(1)**, **1637(m)**

b) Platforms must be planked solid (without gaps) and cover the entire space between scaffold uprights. **1640(b)**, **1641(g)**, **1644(a)**, **1646(e)**

*Exception: In solid planking the following gaps are permissible:*

(1) The opening under the back railing:

- Wood scaffolds: 8 in. (max) horizontal. **1640(b)(5)**

- Metal scaffolds: 10 in. (max) horizontal. **1644(a)(7)**

(2) Space between the building (structure) and the platform:

- Wood scaffolds: 14 in.(max). **1640(b)(5)**

- Metal scaffolds: 16 in. (max). **1644(a)(7)**

- Bricklayers scaffolds: 7 in. (max) to finished face of building. **1641(g)(2)**

c) Platform minimum widths are as follows:

(1) Light duty: 20 in. **1640(b)(5)**

(2) Heavy trades: 4 ft. **1641(c)**

d) Platform slope must not exceed 2 ft. vertically to 10 ft. horizontally. **1637(o)**

e) Overhead protection is required when people are working overhead. **1637(q)**

f) Slippery platform conditions are prohibited. **1637(p)**

g) All scaffold platforms shall meet the planking requirements of Section **1637**. **3622(f)(5)**

9. Planking must conform as follows:

a) All solid sawn planking, unless specified in other orders, must be made of scaffold grade (structural plank 2200 Psi) lumber (see **1504**) with a nominal dimension of at least 2" x 10". **1637(f)(1)**

Prior to being placed into service, all solid sawn wood scaffold planks shall be certified by, or bear the grade stamp of, a grading agency approved by the American Lumber Standards Committee. **1637(f)(5)**

b) All Douglas Fir and Southern Pine planking sized 2 x 10-inch (nominal) or 2 x 9-inch (rough) shall not exceed a maximum span as follows: **1637(f)(2)**

(1) Light trades @ 25 psf = 10 ft.

(2) Medium trades @ 50 psf = 8 ft.

(3) Heavy trades @ 75 psf = 7 ft.

c) The maximum permissible spans allowed for other wood species of scaffold planking shall not exceed 10 feet and shall be determined by a licensed professional engineer. **1637(f)(3)**

(1) All manufactured scaffold planking including engineered wood products, laminated veneer lumber, metal, composite, plastic planks shall be capable of supporting, without failure, its own weight and 4 times the maximum intended working load.

(2) Prior to being placed in service, all laminated veneer lumber scaffold planks, manufactured after December 2, 2010 shall be labeled with the seal of an independent, nationally recognized, inspection agency approved by the International Accreditation Services (IAS) certifying compliance with ASTM D 5456-09a and ANSI/ASSE A10.8-2001, Section 5.2.10.

(3) Planks with spans in excess of 10 feet shall be labeled to indicate the maximum intended working load.

(4) Planks shall be used in accordance with the manufacturer's specifications.

d) All scaffold planks shall be visually inspected for defects before use each day. **1637(f)(6)**

e) Defective or damaged scaffold planks shall not be used and shall be removed from service. **1637(f)(7)**

f) Planking shall overhang the ledger or support as follows:

(1) A minimum of 6 in. **1640(b)**, **1645(b)**

(2) A maximum of 18 in. **1637(g)**, **1645(b)**

g) A single plank (up to 4 ft. high) is only permitted on light-trade wooden pole and horse scaffolds. **1640(b) (5)(A)**, **1647(e)(2)**

h) All platform planks, shall not deflect more than 1/60 of the span when loaded to the manufacturer's recommended maximum load. **1637(w)**

10. Guardrails must be installed on open sides and ends of platforms that are 7 1/2 ft. or higher. **1621(a)**

*Exception:* **1644(a)(6)(A),(B)**

» X braces that substitute for a midrail must intersect *20 in. to 30 in. above the platform.*

» *X-braces that substitute for a top rail must intersect 42 in. to 48 in. above the platform, and a midrail must be placed at 19 in. to 25 in. above the platform.*

11. Toeboards are required on all railed sides of work surfaces where employees work or pass below. **1621(b)**

12. Height limits for scaffolding are as follows:

a) Wood (frame/post) = 60 ft. **1643**

b) Tube and coupler = 125 ft. **1644(b)(4)**

c) Tubular (welded) = 125 ft. **1644(c)(7)**

*Exception: The above limits do not apply when the scaffolding is designed by a civil engineer registered in California.*

    d) Horse (single) = l0 ft. **1647(b)(2)**

    e) Horse (tiered) = 10 ft. **1647(b)(2)**

13. Prohibited scaffolds and supports: **1637(j)**

    a) Shore scaffolds

    b) Jack scaffolds (with brackets attached to single studs)

    c) Lean-to scaffolds

    d) Stilts

    e) Nailed brackets

    f) Brick or blocks

    g) Loose tile

    h) Unstable objects

14. Maximum scaffold working load must be posted or provided to and available from the jobsite supervisor. **1637(b)(6)**

15. Prohibited work practices:

    a) Work on or from scaffolds during storms or high winds unless: **1637(u)**

      (1) A qualified person has determined that it is safe, and

      (2) Employees are protected by a personal fall arrest system, or wind screens.

*Note: Wind screens shall not be used unless the scaffold is secured against the anticipated wind forces.* **1637(u)**

    b) Wood platforms shall not be painted with opaque finishes, but can be coated with certain clear finishes. **1637(v)**

B. Scaffold-specific requirements

After you have reviewed the general requirements for scaffolds, refer to the regulations listed below (and any other applicable SO's ) for the specific type(s) of scaffold you are using to determine whether these requirements replace or augment the general requirements.

The requirements listed below are unique to each specific type of scaffold listed:

1.  Tubular welded scaffold systems. **1644**
    These scaffold systems are commercially fabricated and must meet the following requirements:

    a) Frames must nest with coupling or stacking pins to provide proper vertical alignment. **1644(c)(5)**

    b) Frame panels must be vertically pinned if uplift may occur. **1644(c)(6)**

2.  Tower and rolling scaffolds. **1646**
    The specifications for tower and rolling scaffolds are as follows:

    a) The "height-to-base" must not exceed 3:1 unless the scaffold is secured. **1646(a)**

    b) A screw jack must extend 1/3 of its length into the leg tube, and the exposed thread must not exceed 12 in. **1646(b)(2)**

    c) Two wheels, or casters, must swivel; all four must lock. **1646(c)**

    d) A fully planked platform is required. **1646(e)**

    e) All frame and center joints shall be locked together by lock pins, bolts, or equivalent fastenings. **1646(d)**

    f) The scaffold must have horizontal diagonal bracing (see Illustration 9). **1646(b)**

    g) Railings are required if the platform is 7 1/2 ft. or more above grade. **1646(b)**

h) Ladders or other unstable objects shall not be placed on top of rolling scaffolds to gain greater height. **1646(f)**

i) When scaffolds are built on motor trucks or vehicles, they must be rigidly attached to the truck or vehicle. **1646(g)**

j) Trucks or vehicles that have scaffolds attached to them shall have a device in use whenever employees are on the scaffold that prevents swaying or listing of the platforms. **1646(h)**

k) Employees may ride on rolling scaffold moved by others below if the following conditions exist: **1646(i)**

   (1) The floor or surface is within 3 degrees of level, and free from pits, holes, or obstructions.

   (2) The minimum dimension of the scaffold base, when ready for rolling, is at least 1/2 of the height. Outriggers, if used, shall be installed on both sides of staging.

   (3) The wheels are equipped with rubber or similar resilient tires. For towers 50 feet or over, metal wheels may be used.

   (4) The manual force used to move the scaffold shall be applied as close to the base as practicable, but not more than 5 feet (1.5 meters) above the supporting surface of the scaffold.

   (5) Before a scaffold is moved, each employee on the scaffold shall be made aware of the move.

   (6) No employee shall be on any part of the scaffold which extends outward beyond the wheels, casters, or other supports.

l) Employees may ride and move on a self-propelled rolling scaffold while on the platform without assistance from others below provided the following conditions are met: **1646(j)**

    (1) All of the provisions in **1646(i)** shall be met, except that the scaffold need not be moved by others below.

    (2) The scaffold platform shall not be more than 4 feet above the floor level.

    (3) The working platform shall be no less than 20 in. in width with a maximum 1 inch space between platform planks.

    (4) Wheels or casters of rolling scaffolds shall be provided with an effective locking device that is used in accordance with **1646(c)** or rolling scaffolds shall be provided with an effective device that is used to prevent movement of the scaffold when workers are climbing or working on the scaffold.

    (5) The use of power systems such as motor vehicles, add-on motors, or battery powered equipment to propel a rolling scaffold is prohibited.

m) Employees who ride on rolling scaffolds and employees that assist in moving employees riding on a rolling scaffold shall be trained on the hazards associated with riding on a rolling scaffold as per **1646** and **1509**.

**Illustration 9**
**Tower and Rolling Scaffold**

Working platform

Guardrails

Guardrail support

Access ladder

Locking pins

Toeboard

Locking casters

Cross-bracing

Horizontal diagonal brace

3. Suspended Scaffolds: **1658**

a) General requirements for suspended scaffolds (swing staging). **1658**

Most suspended scaffolding has a two-point suspension supported by hangers or stirrups. The following applies:

(1) Each wire is suspended from a separate outrigger beam or thrustout. **1658(k)**

(2) Multi-stage units or units with overhead protection must be equipped with additional suspension lines to support the scaffolding in case the primary suspension system fails. **1658(u)**

(3) The scaffold must be inspected daily by a qualified person and tested frequently. **1658(g)**

(4) When a suspended scaffold is left unattended in an elevated position, it shall be securely lashed to the building and be cleared of all tools, buckets, or other moveable materials. **1658(p)**

(5) All hoisting mechanisms and metal platforms must meet nationally recognized standards. **1658(a)**

(6) Outrigger beams must be secured in a saddle and anchored at one end to solid structure. The inboard end must be tied back. **1658(j)**

(7) The beam must be capable of supporting four times the intended load. **1658(j)(1)**

(8) Use of a ladder as a platform is prohibited even if a horizontal work surface is added over the rungs. **1658(d)**

(9) The load limit is one person per suspension rope. **1660(a)**

(10) An insulated wire suspension rope is required when workers are welding, burning, sandblasting, or using any chemical substance which may damage the rope. **1658(f)**

(11) A separate safety harness and lifeline are required for each worker. **1658(i)**, **1660(g)**

(12) Platform dimensions must be as follows:

   » Width = 14 in. to 36 in. 1660(d) = 24 in. to 36 in. if the platform is used by cement masons. **1661(b)**

   » Span = 10 ft. (2" x 10" planks). 1660(e) = 12 ft. (2" x 12" planks). **1660(e)**

   » Bolster (ledger) = 2" x 4" cross section. **1660(c)**

b) Specific requirements for suspended scaffolds:

(1) Powered suspended scaffolds. **1667**

The general rules for swing scaffolds apply except as listed below:

- The minimum platform width must be 20 in. **1667(d)**

- Railings are required on open sides and ends and on all sides if the scaffold is suspended by one rope. **1667(a)**

- The load limit is 425 lbs. for a ladder-type platform. **1667(b)**

- Controls must be of the dead-man type.

- Load release units for fast descent are prohibited. **1667(f)(1)**

(2) Interior hung suspended scaffolds. **1665**

These scaffolds are of a wood- or steel-tube-and-coupler type, and they are suspended from a ceiling or roof structure. The general and suspended scaffold rules apply.

Exception:

» *Suspension ropes must be wrapped twice around supporting members and ledgers.* **1665(b)**

» *Ends of wire rope must be secured with at least three clips.*

(3) Float suspended scaffolds. **1663**

These scaffolds are intended for such work as welding, riveting, and bolting. **1663(a)**

» Platform size: 3 ft. x 6 ft. x 3/4 in. plywood. **1663(a)(1)**

» Rope: 1-in. diameter manila (min.). **1663(a)(4)**

» Load limit: three people. **1663(a)**

» Personal fall protection and a separate lifeline: required for each person. **1663(a)(5)**

(4) Boatswain's chair. **1662**

The use of a boatswain's chair requires training or experience. **1662(a)**

» Platform size: 10 in. x 24 in. x 2 in. **1662(i)**

» Rope: 5/8-in. diameter manila (min.) and 3/8-in. diameter protected wire for welding. **1662(j),(k)**

» Personal fall protection and a separate lifeline required. **1662(c)**

» Area below: barricaded. **1662(b)**

(5) Needle Beam scaffolds. **1664**

The specifications for needle beam scaffolds are as follows:

» Beam size: 4 in. x 6 in. x 10 ft. **1664(a)(1)**

» Rope: 1 1/4-in. diameter manila **1664(a)(4)**

» Personal fall protection required in accordance with Article 24 in the CSOs. **1664(a)(12)**

Note: See the hitches for holding needle beams in Illustration 10.

## Illustration 10
### Hitches for Holding Needle Beams

Square knot     Bowline     Rolling or taut-line hitch

Scaffold hitch     Clove hitch     Round turn and two half-hitches

Eye splice     Running bowline     Round turn and two half-hitches

(6) Outrigger scaffolds. **1645**

Outrigger scaffolds are regulated as follows:

» Brackets or beams must be anchored or braced against turning, twisting, or tipping. **1645(a)(1)**

» Platform: at least two 2 in. x 10 in. planks. **1645(a)(2), 1645(b)(5)**

» Beam size: 3 in. x 12 in. (min.) **1645(a)(2)**

» Beam length: Outboard of fulcrum must not exceed 6 ft; inboard must be 1 1/2 times the outboard section. **1645(a)(1)**

Note: For multi-level structures the units must be designed by a California registered Civil Engineer. **1645(a)(3)**

(7) Bracket scaffolds (light trades). **1645**

Brackets must be bolted through walls, welded to tanks, properly secured to metal studs, or hooked over a supporting member. **1645(d)**

» Platform: 20 in. x 10 ft. (min.)

» Load limit: carpenter's type = two workers and 75 lbs. of equipment. **1645(e)(4)**

(8) Horse scaffolds. **1647**

The specifications for horse scaffolds are as follows:

» Platform width:

   i.   Light trades = 20 in. (min.); 10 in. if the platform is less than 4 ft. high.

   ii.  Heavy trades = 4 ft. (min.) **1647(e)(2)**

» Width of base legs = 1/2 x height (min.) **1647(a)(3)**

» Height:

   i.   Collapsible horse = 6 ft. (max.) **1647(d)(2)**

   ii.  Single horse = 10 ft. (max.) **1647(e)(1)**

   iii. Two tiers (max.) = 10 ft. (max.) **1647(e)(1)**

(9) Ladder jack scaffolds. **1648**

The specifications for ladder jack scaffold platforms are as follows:

» Span = 16 ft. (max.) **1648(b)**

» Height = 16 ft. (max.) **1648(a)**

» Width = 14 in. (min.) **1648(b)**

» Load = two workers (max.) 1648(a)

Notes:

  » Ladders must Type I, IA, or IAA duty rated ladders in accordance with *3276(c).* Job-built ladders shall not be used for this purpose. *1648(d)*

  » A safety line is required for each worker. **1648(c)**

(10) Window jack scaffolds. **1654**

  The specifications for window jack scaffolds are as follows:

  » Only one window per scaffold is permitted. **1654(d)**

  » The load limit is one person per scaffold. **1654(d)**

  » Fall protection or railings are required. **1654(c)**

## Silica Dust

Construction work that involves exposure to crystalline silica containing materials can cause lung diseases These silica containing materials include (but are not limited to):

  » Airborne sand

  » Rock

  » Ceramic and terracotta tiles

  » Concrete and concrete block

  » Manufactured stone

  » Roof tiles

  » Bricks and blocks

  » Grouts and mortar

  » Some joint compounds

  » Abrasive materials

Exposure to crystalline silica can cause a variety of lung diseases including silicosis, lung cancer, COPD (chronic obstructive pulmonary disease), decreased lung function and increased likelihood of getting tuberculosis. Although most cases of silicosis develop after years of exposure, instances of extremely high exposure have resulted in illness and even death in a matter of weeks. Airborne permissible exposure limits (PELs) are established for several different forms of crystalline silica. These limits range from 0.05 to 0.1 mg/m3 of respirable dust, expressed as an 8hour TWA (see Table AC-1 of **5155**).

**Toolbox**
"Exposure to Respirable Silica"
**PUB223**   www.oshatools.com

Hazardous work activities include abrasive blasting with sand and loading, dumping, chipping, hammering, cutting, and drilling of rock, sand, or concrete. Generally during work on materials, such as rock or concrete, that contain a significant amount of silica (20% or greater), continuous exposure to a visible cloud of dust will probably result in levels of exposure that exceed the PELs. However, in some cases the PELs can be exceeded even when there is no visible cloud of dust.

Before beginning work that could expose employees to crystalline silica, employers must comply with the following requirements:

A. Employers must monitor and control employees' exposure to airborne contaminants. **5155(c),(e), 1530**

B. During operations in which powered tools or equipment are used to cut, grind, core, or drill rock, concrete or masonry materials, a dust reduction system shall be applied to control of employee exposures to airborne particulate. For exceptions, see **1530.1**.

C. Operations in which employees may be repeatedly exposed to rock dust or sand should be evaluated by an individual competent in industrial hygiene practice. Assistance can also be obtained from the Cal/OSHA Consultation Service.

> **Toolbox**
>
> "Silica Hazard Alert"
>
> **PUB265** www.oshatools.com

D. Employers must train supervisors and employees prior to their job assignments. The training shall be provided at least annually and include, but not be limited to, the following: **1530.1(e)**

1. Safety and health hazards of silica dust overexposure.

2. Methods used by the employer to control employee exposures to airborne silica dust.

3. Proper use and maintenance of dust reduction systems.

4. The importance of good personal hygiene and housekeeping practices.

5. Proper use of respirators when required. **5144**, **5194**

## Stairways

Stairways are an acceptable method for gaining access to floors and working levels of buildings and scaffolds.

In addition to the stairways required, buildings 60 ft. or more in height or 48 ft. below ground level require an elevator. **1630(a)**

Stairways must be installed as follows:

A. In buildings of up to three stories or 36 ft. in height, at least one stairway is required. **1629(a)(4)**

B. In buildings of more than three stories or 36 ft. in height, two or more stairways are required. **1629(a)(4)**

C. A stairway to a second or higher floor must be installed before studs are raised to support the next higher floor. **1629(b)(1)(A)**

D. In steel frame buildings, a stairway must be installed leading up to each planked floor. **1629(b)(2)**

E. In concrete buildings, a stairway must be installed to the floor that supports the vertical shoring system. **1629(b)(3)**

F. Stairways shall be at least 24 in. in width and shall be equipped with stair rails, handrails, treads, and landings.

G. All guardrails railings, including their connections and anchorage, shall be capable of withstanding a load as specified in **1620(c)**.

H. Handrails must be 34 in. to 38 in. above the tread nosing. **1626(c)(6)**

I. Wooden posts shall be not less than 2 in. by 4 in. in cross section, spaced at 8-foot or closer intervals. Wooden top railings shall be smooth and of 2-in. by 4-in. or larger material. Double, 1-in. by 4-in. members may be used as top railings when certain conditions are met. **1620(b)(2),(3)**

J. Railings and toeboards must be installed around stairwells. **1626(a)(2)**

K. The stairway shall have landings at each floor, or level, of not less than 30 in. in the direction of travel and extend at least 24 in. in width at every 12 feet or less of vertical rise. **1626(a)(2)**

L. Stair steps must be illuminated with at least 5-ft. candles of light and all lamps must be guarded. **1629(a)(7)**

## Toeboards

Regulations concerning toeboards include the following:

A. Toeboards must be provided on all open sides and ends of railed scaffolds at locations where persons are required to work or to pass under the scaffold and at all interior floor, roof, and shaft openings. **1621(b)**

B. Specifications for toeboards are as follows:

   1. A toeboard must be securely fastened at a minimum of 4 in. (nominal) in height from its top edge to the level of the floor, platform, runway, or ramp. A toeboard must have not more than a 1/4-in. clearance above the floor level. It may be made of any substantial material, either solid, or with openings not more than 1 in. in greatest dimension. **1621(b)**

   2. Where material is piled to such a height that a standard toeboard does not provide protection, paneling or screening from floor to intermediate rail or top rail shall be provided. **1621(c)**

## Toilets/Washing Facilities/Sanitation

Regulations concerning toilets, hand washing, and sanitation include the following:

A. Toilet facilities are required at the job site. **1526(b)**

B. A toilet is required for each 20 employees or fraction thereof for each sex; urinals may be substituted for half of the units. **1526(a)**

   *Exception: Sites with fewer than five employees are not required to provide separate toilets for each sex; however, toilets must be lockable from the inside. 1526(a)*

C. Toilets must be kept clean and supplied with toilet paper. **1526(d)**

D. Toilets are not required for mobile crews if transportation to nearby toilets is available. **1526(e)**

E. One washing station must be provided for each 20 employees or fraction thereof. **1527(a)**

F. Washing stations must be clean and have an adequate supply of soap, water, and single use towels (or warm air blower). **1527(a)**

G. Washing station must have a sign indicating water is for washing. **1527(a)(1)(F)**

H. Wash stations are to be located outside and not attached to the toilet facility. **1527(a)(1)(F)**

*Exception: Where there are less than 5 employees and only one toilet facility is required, the wash station may be located inside the toilet facility.*

I. If showering is required by the employer, the shower must meet specific requirements. **1527(a)(3)**

J. An adequate supply of potable (drinkable) water must be provided at each job site. The employer shall take one or more of the following steps to ensure every employee has access to drinking water: **1524(a)**

1. Provide drinking fountains.

2. Supply single-service cups.

3. Supply sealed one-time use water containers.

4. Ensure re-usable, closable containers are available for individual employee use.

*Note: Additional requirements for the provision of drinking water in outdoor places of employment are contained in* **3395***.*

## Tools and Equipment

General Requirements for Tools and Equipment Include:

» Tools must be kept clean and in good repair. **1699**

» Only trained or experienced employees may operate tools, machines, or equipment. **1510(b)**

» Power-operated tools must be grounded or of the double-insulated type. If double-insulated types or tools are used, the equipment shall be distinctively marked. **2395.45**

» Power-operated tools should be kept out of wet locations. **2395.45**

A. Power-operated tools shall be grounded under the following conditions: **2395.45**

1. Utilization equipment used in hazardous (classified) locations. (**See Article 59**)

2. Hand-held motor-operated tools, stationary and fixed motor-operated tools, and light industrial motor-operated tools.

3. Motor-operated tools and utilization equipment of the following types: drills, hedge clippers, lawn mowers, snow blowers, wet scrubbers, sanders and saws.

4. Tools likely to be used in wet and conductive locations.

**Toolbox**
"Nail Gun Safety"
**PUB212** www.oshatools.com

*Notes:*

» *The followings shall not be required to be grounded.* **2395.45**

    *i. Listed portable tools or utilization equipment likely to be used in wet and conductive locations if supplied through an isolating transformer with an ungrounded secondary of not over 50 volts.*

    *ii. Listed or labeled portable tools and utilization equipment protected by an approved system of double insulation. Where such a system is employed, the equipment shall be distinctively marked.*

» Double-insulated type power-operated tools are not required to be grounded.

B. Guards required by the SOs must not be removed or deactivated. **3942**

C. Control switches for powered hand tools are subject to the regulations noted below:

1. The following tools must be equipped with a constant-contact (dead-man) on-off switch: **3557(a)**

    a) Drills

    b) Tappers

    c) Fastener drivers

    d) Grinders

    e) Disc and belt sanders

    f) Reciprocating saws

    g) Circular saws

    h) Chain saws

    i) Concrete vibrators

    j) Concrete breakers

    k) Concrete trowels

l) Powered tampers

m) Jack hammers

n) Rock drills

o) Tools similar to those above

2. Hoisting or lowering electric tools by their cords is prohibited. **1707(a)**

D. Powder-actuated tools (PAT) shall be approved for their intended use as defined in **1505**, or have California approval numbers. **1684(a)(1),(2)**

1. Only trained workers holding a valid operator's card may use a PAT. **1685(a)(1)**

2. Containers must be lockable and bear a label that says POWDER-ACTUATED TOOL on the outside. The storage container must be kept under lock and key. **1687(a)**

3. The PAT must be provided with the following:

   a) An operating and service manual.

   b) A power load and fastener chart.

   c) An inspection and service record.

   d) Repair and servicing tools. **1687(b)**

4. Limitations on the use of PATs are as follows:

   a) Workers must not leave the tool unattended. **1690(b)**

   b) Workers must not use the tool:

   (1) In an explosive environment. **1690(a)**

   (2) On hard or brittle material. **1690(c)**

   (3) On easily penetrated or thin materials or materials of questionable resistance unless backed. **1690(d)**

   (4) Within a 1/2 in. of the edge of steel. **1690(e)**

(5) Within 3 in. of the edge of masonry. **1690(f)**

(6) On thin concrete. **1690(g)**

(7) On spalled areas. **1690(h)**

(8) On existing holes. **1690(i)**

5. Requirements for operating PATs are as noted:

a) Eye or face protection is required for operators and assistants. **1691(b)**

b) Operators must inspect the tool before using it. **1691(c)**

c) Defective tools must not be used. **1691(d)**

d) Tools must not be loaded until ready for use. **1691(g)**

e) Tools must be unloaded if work is interrupted. **1691(h)**

f) Operators must never point a loaded tool or an empty tool at anyone. **1691(i)**

g) The tool must be held in place for 30 seconds on misfire. **1691(l)**

h) Different power loads must be kept in separate compartments. **1691(m)**

i) Warning signs that say POWDER-ACTUATED TOOLS IN USE must be conspicuously displayed within 50 ft. of a PAT operation. **1691(n)**

j) Misfires and skipped power charges must be stored and disposed of properly. **1689(c)**, **1691(a)**

E. Concrete-finishing tools must be equipped with a dead-man-type control. **1698(d)**

F. Airless spray guns must have an automatic or visible manual release safety device or a diffuser nut and tip guard. **3559.1(a)**

G. Circular power saws are regulated as follows:

1. Portable Circular power saws:

   a) Teeth on the upper half of the saw blade must be permanently guarded. **4307(a)**

   b) Teeth on the lower half of the saw blade must be guarded with a telescopic or hinged guard. **4307(b)**

   c) Saw guards must not be blocked open to prevent guards from functioning. **4307(c)**

2. Self-feed Circular power saws: **4301**

   a) In addition to guards over blades as specified in **4296,** feed rolls shall be protected by a hood or guard.

   b) The employer shall ensure that power feed devices are properly adjusted for each piece of stock in order to reduce the possibility of kickback.

   c) Every self-feed circular ripsaw shall be equipped with an anti-kickback device installed on the infeed side.

*Note: The arbor speed of circular saw blades shall not exceed speeds recommended by the manufacturer.*

H. Miter (chop) saws are regulated as follows: **4307.1**

1. With the carriage in the full cut position, a guard must enclose the upper half of the blade and at least 50 percent of the arbor end. **4307.1(a)**

2. With the carriage in the full retract (raised) position, lower blade teeth must be fully guarded, and the guard must extend at least 3/4 in. beyond the teeth. **4307.1(b)**

3. Employers shall instruct employees to keep hands and fingers outside the area below the blade until the blade has come to a complete stop. **4307.1(c)**

I. Radial arm (horizontal pull) saws are regulated as follows:

1. The upper half of the saw blade and arbor ends must be completely covered. **4309(a)**

2. An anti-kickback device must be used during ripping operations. **4309(c)**

3. Saws must return automatically to the table's back when released. **4309(d)**

4. Saws must have a stop provided to prevent the saw blade from passing the front edge of the table. **4309(b)**

J. Table saws are regulated as follows:

1. A hood must cover the saw to at least the depth of the teeth. **4300(a)**

2. The hood shall automatically adjust itself to the thickness of the material being cut at the point where the stock encounters the saw blade. The hood may be a fixed or manually adjusted hood or guard provided the space between the bottom of the guard and the material being cut does not exceed 1/4 inch. **4300(b),(c)**

3. Table saws must be equipped with an anti-kickback device during ripping operations. **4300(d)**

4. Push sticks or push blocks shall be provided at the work place in the several sizes and types suitable for the work to be done. **4300(f)**

*Note: The arbor speed of circular saw blades shall not exceed speeds recommended by the manufacturer.*

K. Band saws are regulated as follows:

1. All portions of the band saw blade must be guarded except between the guide rolls and the table. **4310(a)(1)**

2. Band saw wheels must be enclosed. **4310(a)(2)**

L. Chain saws are regulated as follows:

1. Chain saws must be equipped with a constant-pressure control that returns the saw to idling speed when released. **3425(a)(2)**

2. Chain saws must have a clutch adjusted to prevent the chain drive from engaging at idling speed. **3425(a)(3)**

M. Pneumatic tools are regulated as follows:

1. Safety clips are required on pneumatic tools to prevent dies from being accidentally expelled from the barrel. **3559(a)**

2. Pneumatic nailers and staplers must have a safety device that prevents the tool from operating when the muzzle is not in contact with the work surface. **1704(b)**

*Exception: Light-Duty Nailers and Staplers*

3. Pneumatic nailers and staplers must be disconnected from the air supply at the tool when performing any maintenance or repair on the tool, or clearing a jam. **1704(c)**

4. The air hose of pneumatic nailers and staplers must be secured at roof level to provide ample but not excessive amounts of hose when an operator works on roofs sloped steeper than 7:12. **1704(d)**

5. All pneumatic hoses exceeding 1/2-inch inside diameter shall have a safety device at the source of supply or branch line to reduce pressure in case of hose failure. **1704(e)**

6. Jack hammer operators must wear personal protective equipment when required (see Personal Protective Equipment section in this guide), including foot protection as per **3385**. Jack hammer operators must also use hearing protection when noise levels exceed allowable exposure levels as per **5096(a)**.

N. All portable pipe threading/cutting machines, portable power driven augers (earth drills), and portable power drives shall be permanently equipped with a momentary contact device. **4086**

# Traffic Control

Regulations concerning traffic control are noted below:

A. Worksite traffic controls and placement of warning signs must conform to the requirements of the "California Manual on Uniform Traffic Control Devices for Streets and Highways," published by the State Department of Transportation. Additional means of traffic control, such as continuous patrol, detours, barricades, or other techniques for the safety of employees may be employed. **1598(a)**

> **Toolbox**
> "Building Safer Highway Work Zones"
> **PUB218**  www.oshatools.com

B. Specifications for the size and design of signs, lights, and devices used for traffic control shall be as described in the "Manual", pursuant to the provisions of California Vehicle Code Section 21400, which is incorporated by this reference. **1598(b)**

> **Toolbox**
> "Work Zone Operations, Best Practices"
> **PUB246**  www.oshatools.com

C. Employees (on foot), such as grade-checkers, surveyors and others exposed to the hazard of vehicular traffic, shall wear high visibility safety apparel in accordance with the requirements of **1598** and **1599. 1590**

*Note: The warning garments such as vests, jackets, or shirts shall be manufactured in accordance with the requirements of the ANSI/ ISEA 107-2004, High Visibility Safety Apparel and Headwear. 1598(c)*

D. Flaggers (see Flaggers section in this guide) are required when the controls cited above are inadequate. **1599(a)**

*Note: The use of one flagger under specified circumstances is also permitted.* **1599(a)**

E. The employer shall select the proper type (class) of high visibility safety apparel for a given occupational activity by consulting the Manual, apparel manufacturer, ANSI/ISEA 107-2004, Appendix B or the American Traffic Safety Services Association (ATSSA). **1599(f)**

> ### Toolbox
> "California Manual on Uniform Traffic Control Devices, Temporary Traffic Control, Part 1&6"
> **PUB247**  www.oshatools.com

## Training

Each year serious and fatal injuries are caused by ineffective and inadequate training of employees. Employees who are newly hired, given new job duties, or who are using tools and equipment that they are unfamiliar with have a greater risk of being injured.

A. Effective Training

Effective training relates directly to the work being done by employees. It instructs employees on general safe work practices and also provides specific information on the hazards they may encounter in their jobs. In general effective training instructs employees on how to work safely, and:

1. Communicates information in a language and by methods understandable to all employees (including those who do not speak English, or have limited comprehension of English).

> ### Toolbox
> "Safety and Health Training Requirements"
> **PUB264**  www.oshatools.com

2. Helps establish a relationship with employees to improve trust and communication.

3. Is participatory and involves employees by drawing on their own real life experiences.

4. Allows group hazard identification and problem solving by means of demonstrations, asking questions, discussing ideas, and providing observations and stories.

5. Provides opportunities to demonstrate newly learned safe work practices and the safe use of tools, equipment, and chemicals.

6. Provides concrete safety and health changes in how work is set-up and performed.

7. Is repeated as often as necessary.

8. Encourages employees to express safety concerns and to make suggestions.

> **Toolbox**
> "Strategies for Improving Worker Training"
> **PUB233**   www.oshatools.com

B. Training Requirements

The specific Cal/OSHA training requirements that apply to each worksite depend on the work activities that employees are involved in.

The SOs require training employees when:

1. They are first hired. **1510(a), 3203(a)**

2. They will operate machinery and equipment (see the Qualified Person section in this guide).

3. They are given a new job assignment for which they have not previously received training. **3203(a)(7)(C)**

4. They are exposed to known job-site hazards, such as poisons, hazardous materials and gases, harmful plants and animals, etc. **1510(c)**

5. New substances, processes, procedures, or equipment are introduced to the workplace and represent a new hazard. **3203(a)(7)(D)**

6. The employer is made aware of a new or previously unrecognized hazard. **3203(a)(7)(E)**

7. Supervisors need to familiarize themselves with the safety and health hazards to which employees under their immediate direction and control may be exposed. **3203(a)(7)(F)**

8. Tailgate or toolbox safety meeting are held (at least every ten working days). **1509(e)**

   *Exception: For tunneling operations tailgate meetings must be held weekly.* **8406(e)**

*Note: Cal/OSHA has a large number of regulations which require employee training. The list above includes only some of the Cal/OSHA regulations which require training.*

## Tunnels and Tunneling

Employees working on tunneling operations are exposed to numerous hazards, including (1) tunnel collapses; (2) hazardous atmospheres; and (3) explosive atmospheres. Employees working on pressurized tunneling operations may also be exposed to hazardous hyperbaric conditions.

When employees work in tunnels and in underground chambers of any depth and in shafts planned to exceed 20 ft. in depth, the following operations are subject to the Tunnel Safety Orders (TSOs):

» Pipe-jacking and boring

» Micro-tunneling

» Mechanized tunneling

» Drill and blast work

» Excavation

» Ground support work

» Repair and maintenance

» Tunnel renovations

Employees who are exposed to or control the exposure of others to hyperbaric conditions shall be trained in hyperbaric related physics and physiology, recognition of pressure related injuries, and how to avoid discomfort during compression. **6075(c)**

The Mining and Tunneling (M&T) Unit of Cal/OSHA enforces the TSOs, which include:

A. Classifications: The M&T Unit is required to classify the gas hazards of each tunnel or shaft. These classifications are (1) nongassy; (2) potentially gassy; (3) gassy; and (4) extra hazardous. **8422 (a),(b)**

   Note: The request for classification shall be sent to the nearest M&T Unit office.

B. Pre-job safety conference: Before underground excavation may begin, the M&T Unit must conduct an on-site, pre-job safety conference with the project owner, the general contractor, the tunnel contractor, and the tunnel contractor's employees. The goal of the conference is to ensure that all of the employees are aware of the conditions under which the tunnel will be driven and that all of the safety issues are discussed and problems resolved. **8408**

C. Certified persons: Cal/OSHA requires the persons performing the duties of gas tester or safety representative to be certified by passing a written and an oral examination administered by the M&T Unit. **8406(f),(h)**

   1. A certified gas tester is required for the following operations:

      a) After blasting operations.

      b) Projects during which diesel equipment is used underground.

      c) Hazardous underground gas conditions. **8406**

2. A certified safety representative must direct the required safety and health program and must be on-site while employees are engaged in operations during which the TSOs apply. The safety representative must have knowledge in underground safety, must be able to recognize hazards, and must have the authority to correct unsafe conditions and procedures subject to the TSOs. **8406(f)**

D. Diesel engines: Diesel engines are the only type of internal combustion engine acceptable for use during tunneling operations, provided that the following requirements are met:

1. Cal/OSHA must issue a permit for diesel engine operation.

2. Conditions of the permit must be observed.

3. Ventilation and fresh air flow must meet the required minimum standards.

4. Air concentrations of nitrogen dioxide, carbon monoxide, and carbon dioxide in the tunnel must be determined at least once during each shift at the peak of diesel operation and kept at or below the PELs.

5. A written record must be kept of the above readings.

6. PELs of the above air contaminants or any other contaminants must not be exceeded.

7. A certified gas tester must conduct the testing (see additional requirements in **8470**).

8. An approved exhaust purifier must be installed and maintained (see the requirements in **8470**).

E. Licensed blasters: All blasting at tunnel sites shall be carried out or directly supervised onsite by a California licensed blaster as required by TSO **8560**.

# Welding, Cutting, and Other Hot Work

Each year numerous deaths from explosions, electrocutions, asphyxiation, falls, and crushing injuries are associated with hot work activities. These deaths from hot work often occur in confined or restricted spaces. In addition, numerous health hazards including heavy metal poisoning, lung cancer, metal fume fever, flash burns, and welders flash (burn to the eyes) are associated with exposure to fumes, gases, and ionizing and non-ionizing radiation formed or released during welding, cutting, brazing, and other hot work.

A. Before workers begin a hot work, the following controls must be established:

   1. No welding is permitted in an explosive environment. **4848**

   2. A written "hot work" permit is recommended whenever a combustible environment may exist. **4848**

   3. All combustible materials in the work area must be removed or shielded. **4848**

   4. Suitable fire extinguishers, that meet NFPA and ANSI Standards, must be provided in the work area. **4848**

   5. Welding blankets, curtains and pads shall be approved for their intended use in accordance with Section **3206** of these Orders. **4848(b)**

   6. Employers must instruct employees on hot work safety. **4848(a)**

   7. Welders must be required to wear:

      a) Non-flammable gloves with gauntlets. **1520**

      b) Appropriate foot protection. **3385**

      c) Aprons (leather) and shirts that have sleeves and collars. **1522(a)**

      d) Helmets, hoods, and face shields suitable for head protection. **3381(a), 3382(a)**

e) Suitable eye protection. **3382**

f) Respiratory protection (as required). **5144**

8. Screens must be provided to protect the eyes of nonwelders from flash burns and ultraviolet light rays. **3382(b)**

B. Gas welding is regulated as follows:

1. Fuel gas and oxygen hoses must be distinguished from each other. **1742(a)**

2. Couplings must not disconnect by means of a straight-pull motion. **1742(g)**

3. Oil or grease must never come into contact with oxygen equipment. **1743(c)**

4. Oxygen from a system without a pressure regulation device must never be used. **1743(e)**

5. Gas cylinders must be stored and used as follows:

   a) Cylinders must be protected from all heat sources. **1740(a)**

   b) Cylinders containing oxygen, acetylene or fuel-gases shall not be taken into confined spaces. **1740(b)**

   c) Acetylene and fuel gas cylinders, including but not limited to welding and cutting fuel gas cylinders, shall be stored and used with the valve end up. **1740(b)**

   *Exception: Fuel gas cylinders containing fuel gas used to power industrial trucks regulated by Article 25 of the GISO.*

   d) All gas cylinders in service shall be securely held in substantial fixed or portable racks, or placed so they will not fall or be knocked over. **1740(c)**

   e) Cylinders must be handled in suitable cradles, with their valve caps installed; they must never be lifted by magnet, rope, or chain. **1740(c), (d)**

f) Cylinders must not be placed where they might form a part of any electric circuit. **1740(e)**

g) Oxygen cylinders in storage shall be separated from fuel-gas cylinders or combustible materials (especially oil or grease), a minimum distance of 20 ft. or by a noncombustible barrier at least 5 ft. high having a fire-resistance rating of at least one-half hour. **1740(g)**

h) Valve stem wrenches must be left in place while cylinders are in use. **1743(g)**

i) A fire extinguisher rated at least 10 B:C must be kept near the operation. **1743(j)**

j) Backflow protection is required. **4845(b)**

C. Arc welding is regulated as follows:

1. Cables in poor condition must not be used; no cable may be spliced within 10 ft. of the electrode holder. **4851(e)(2)**

2. The frames of arc welding and cutting machines must be grounded. **4851(f)(5)**

3. Electrodes and holders that are not in use shall be protected so they cannot make electrical contact with employees or conducting objects. **4851(g)**

4. Defective equipment must not be used. **4851(j)**

D. Ventilation regulations for welding, cutting, and brazing operations require that worker's exposure(s) to hazardous fumes, gases, and vapors be reduced below PELs. **1536, 1537, 5155**

1. Outdoor operations
   Respirators are required for any operation involving beryllium, cadmium, lead, or mercury. For other operations and materials, respirators are not required when natural or mechanical ventilation is sufficient to prevent exposure to airborne contaminants in excess of the PELs noted in **5155**. **1536(c)**

2. Indoor operations
   Respirators shall be used when local exhaust or mechanical ventilation is not feasible or able to prevent exposures that exceed limits specified in **5155**.

E. In enclosed spaces, supplied-air respirators shall be used when local exhaust ventilation is not an effective means for preventing potentially hazardous exposures. **1536(b)**, **5152**

## Wood Preservative Chemicals

Wood preservatives that contain creosote, pentachlorophenol, inorganic arsenic, and chromates are widely used. Because these chemicals are carcinogens, exposures to employees must be eliminated or reduced to the lowest levels possible below the PELs by using effective engineering control (for example, enclosure or confinement of the operation, general and local exhaust ventilation, and substitution of less toxic materials). When effective engineering controls are not feasible, or while they are being instituted, use of NIOSH-approved respirators is required to eliminate harmful airborne exposures to these chemicals. **5141**, **5144(a)**, **5214**

When the probability of skin or eye irritation exists, workers must use appropriate protective clothing and equipment, such as coveralls, gloves, shoes, face shields, or impervious clothing.

**Toolbox**
"Pocket Guide to Chemical Hazards"
**PUB228**   www.oshatools.com

**Build Safety Knowledge**

**www.OshaTools.com**

## Appendix

### Safety Publication Toolbox

These safety publications have been job-tested. They cover a wide-range of topics and get the message across.

Consolidated in a central location for easy access, they are available to view and download from www.oshatools.com.

"Cut and paste" the **PUB###** (no spaces) from each publication into the search box at www.oshatools.com.

**BONUS** – www.oshatools.com makes it easy for Supervisors to share safety info with the entire crew. Select the SHARE button located with each publication at www.oshatools.com, add the address, comments and send. No registration and no fee – just easy!

Safety publications are continually being reviewed and added to www.oshatools.com. If you can't find a topic in this Compliance Guide, check www.oshatools.com. The topic you are looking for may have been added to the website.

Some of these publications are written by government agencies other than Cal/OSHA and might not follow Cal/OSHA rules exactly. The reader should be aware of this and use these resources accordingly.

## PUB201

### Aerial Lifts – OSHA

Aerial lifts are creatively used by workers for many tasks. This can cause problems. Many workers are injured or killed on aerial lifts each year. OSHA provides the following information to help employers and workers recognize and avoid safety hazards they may encounter when they use aerial lifts. Aerial lifts have replaced ladders and scaffolding on many job sites due to their mobility and flexibility. They may be made of metal, fiberglass reinforced plastic or other materials. They may be powered or manually operated.

## PUB202

### Asbestos in Construction – OSHA

Asbestos fibers enter the body when a person inhales or ingests airborne particles that become embedded in the tissues of the respiratory or digestive systems. In the construction industry, asbestos is found in installed products such as sprayed-on fireproofing, pipe insulation, floor tiles, cement pipe and sheet, roofing felts and shingles, ceiling tiles, fire-resistant drywall, drywall joint compounds, and acoustical products.

## PUB203

### Concrete Construction – OSHA

I once investigated an injury accident that was a 3rd degree burn experienced by a laborer working in concrete for 8 hours. He had no idea that the extreme alkalinity of the concrete could cause a burn. More than 250,000 people work in concrete manufacturing. Over 10 percent of those workers — 28,000 — experienced a job-related injury or illness and 42 died in just one year.

## PUB204

### Confined Space Entry – OSHA

The terms "permit-required confined space" and "permit space" refer to spaces that meet OSHA's definition of a "confined space" and contain health or safety hazards. For this reason, OSHA requires workers to have a permit to enter these spaces. This publication reviews safe work procedures when working in permitted spaces. I have noticed that the word "permit" sometimes gives a confusing message to workers. Occasionally, I have asked workers to think of the "permit" requirement as actually a "pre-job safety checklist" and the concept seemed easier to grasp.

## PUB205

### Construction Pocket Guide – OSHA

Nearly 6.5 million people work at approximately 252,000 construction sites across the nation on any given day. The fatal injury rate for the construction industry is higher than the national average in this category for all industries. Potential hazards for workers in construction include: Falls (from heights); Trench collapse; Scaffold collapse; Electric shock and arc flash/arc blast; Failure to use proper personal protective equipment; and Repetitive motion injuries.

## PUB206

### Controlling Electrical Hazards – OSHA

It's the volts that jolt, and the mils that kill! Electricity has long been recognized as a serious workplace hazard, exposing employees to electric shock, electrocution, burns, fires, and explosions. Some employees, such as engineers, electricians, electronic technicians, and power line workers, work with electricity directly. Others, such as office workers and sales people, work with it indirectly. Perhaps because it has become such a familiar part of our daily life, we tend to overlook the hazards electricity poses and fail to treat it with the respect it deserves.

## PUB207

### Design of Fire Service Features – OSHA

The purpose of this manual is to increase the safety of building occupants and emergency responders by streamlining fire service interaction with building features and fire protection systems. Architects and engineers create workplaces for firefighters; the information in this manual will assist designers of buildings and fire protection systems to better understand the needs of the fire service when they are called upon to operate in or near the built environment.

## PUB208

### Field Operations Manual – OSHA

OSHA's Inspectors refer to the Field Operations Manual for their inspection policies and procedures. This is a good publication to have in your library, providing a good idea of what action(s) Inspectors will take when they visit your job site. I don't want to imply that you should read from cover-to-cover because it is over 300 pages long; however, Chapter 3, Inspection Procedures, is a must read. This chapter includes how Inspectors choose companies, types of inspections, info on the opening / closing conference and the walk around inspection. Thanks to OSHA for sharing.

## PUB209

### Guide to Hexavalent Chromium Standards – OSHA

Is your business involved with electroplating, welding or painting? This guide is intended to help businesses comply with OSHAs Hexavalent Chromium (Cr(VI)) standards. Employees exposed to Cr(VI) are at increased risk of developing serious adverse health effects including lung cancer, asthma and damage to the nasal passages and skin. Examples of major industrial uses of Cr(VI) compounds include: chromate pigments in dyes, paints, inks, and plastics; chromates added as anticorrosive agents to paints, primers, and other surface coatings; and chromic acid electroplated onto metal parts to provide a decorative or protective coating.

## PUB210

### Heat Illness Prevention – OSHA

Heat exhaustion just doesn't seem real – until you've personally experienced it. Then you respect it. It's mean. The past few summers have shown that the risk of heat illness from high temperatures is one of the most serious challenges to the safety and health of workers. This training guide will help you plan how to prevent heat illness among your crew and provide training to your workers. In addition to safe work checklists and heat illness prevention posters to share with the crew, it also includes training techniques for trainers.

## PUB211

### Lockout/Tagout General Industry – OSHA

Employees can be seriously or fatally injured if machinery they service or maintain unexpectedly energizes, starts up, or releases stored energy. This document addresses practices and procedures necessary to disable machinery and prevent the release of potentially hazardous energy while maintenance or servicing activities are performed. Conduct safety audits of work areas to make sure lockout procedures are done correctly and your employees are protected.

## PUB212

### Nail Gun Safety – OSHA

Nail guns are used every day on many construction jobs— especially in residen¬tial construction. They boost productivity but also cause tens of thousands of painful injuries each year. Nail gun injuries are common—one study found that 2 out of 5 residential carpenter apprentices experienced a nail gun injury over a four-year period. When they do occur, these injuries are often not reported or given any medical treatment. Research has identified the risk factors that make nail gun injuries more likely to occur. The type of trigger system and the extent of training are important factors. The risk of a nail gun injury is twice as high when using a multi-shot contact trigger as when using a single-shot sequential trigger nailer.

## PUB213

### Protecting Yourself from Noise in Construction – OSHA

This is an interesting comment about job site noise - a recent study found that workers persistently exposed to excessive occupational noise may be two-to-three times more likely to suffer from serious heart disease than workers who were not exposed! Loud noise can also reduce work productivity and contribute to workplace accidents by making it difficult to hear warning signals. All valid reasons to reduce job site noise and ensure workers use hearing protection when needed.

## PUB214

### Scaffold Use in the Construction Industry – OSHA

A competent person must inspect the scaffold and scaffold components for visible defects before each work shift and after any occurrence that could affect structural integrity. OSHA's scaffolding standard defines a competent person as "one who is capable of identifying existing and predictable hazards in the surroundings or working conditions, which are unsanitary, hazardous to employees, and who has authorization to take prompt corrective measures to eliminate them." This publication reviews scaffold rules and is a great reference for the Scaffold Competent Person.

## PUB215

### Training Guidelines in OSHA Standards – OSHA

One senior safety professional says he keeps just two books on his desk for reference. One is the OSHA Field Operation Manual and the other is this one, Training Requirements in OSHA Standards. The length and complexity of OSHA standards may make it difficult to find all the references to training. So, to help employers, safety and health professionals, training directors, and others with a need to know, OSHA's training-related requirements have been excerpted and collected in this booklet. Requirements for posting information, warning signs, labels, and the like are excluded, as are most references to the qualifications of people assigned to test workplace conditions or equipment.

## PUB216

### Trench and Excavation – OSHA

Do you know how much one cubic yard of dirt weighs? Over 3000 pounds. There isn't anyone on your crew that can dig or push their way up through 3000 pounds of dirt. No matter how many trenching, shoring, and back-filling jobs you have done in the past, it is important to approach each new job with the utmost care and preparation. Many on-the-job accidents result directly from inadequate initial planning. Waiting until after the work has started to correct mistakes in shoring or sloping slows down the operation, adds to the cost, and increases the possibility of a cave-in or other excavation failure. The Competent Person must be on the job, everyday, all the time. Inspections must be done every shift.

## PUB217

### Analysis of Dozer Accidents – NIOSH

Dozer operator injuries and how they happen. Share with equipment operators. This report describes serious injuries occurring to bulldozer operators working at U.S. coal, metal, and nonmetal mines. The period covered is 1988-97. The data were collected by the Mine Safety and Health Administration (MSHA). A total of 873 injury records are examined. These injuries resulted in 18 fatalities and 31,866 lost workdays. All of these injuries occurred to dozer operators while they were doing common production tasks.

## PUB218

### Building Safer Highway Work Zones – NIOSH

Want to feel uncomfortable? Stand with your road construction crew, 3 feet from live traffic. Highway and street construction workers are at risk of fatal and serious nonfatal injury when working in the vicinity of passing motorists, construction vehicles, and equipment. Each year, more than 100 workers are killed and over 20,000 are injured in the highway and street construction industry. Vehicles and equipment operating in and around the work zone are involved in over half of the worker fatalities in this industry.

**PUB219**

**Compendium of NIOSH Construction Research – NIOSH**

The research program described in this report addresses a variety of important safety and health hazards and conditions. The projects cover the public health spectrum, from identifying and characterizing problems to quantifying and prioritizing risk factors, developing prevention strategies, evaluating results, and disseminating information to construction industry users. Topics include: asphalt, engineering controls, respiratory disease, hearing loss and much more. Projects are described in single page format, by purpose, summary and point of contact.

**PUB220**

**Contacting Overhead Power Lines with Ladders – NIOSH**

LOOK UP! Don't Be Electrocuted! Note the location of overhead power lines at the start of each job. Always assume all overhead lines are energized and dangerous. Do not use metal ladders when working around or near overhead power lines. Always lower the ladder and carry it horizontally when moving it to avoid contacting overhead power lines. Have someone help carry and set up long and unwieldy ladders. Follow the 1:4 rule—For every 4 feet between the ground and the upper point where a ladder is resting, set the feet of the ladder out 1 foot horizontally.

**PUB221**

**Electrical Safety – NIOSH**

Electricity is Dangerous! There are four main types of electrical injuries: electrocution (death due to electrical shock), electrical shock, burns, and falls. The dangers of electricity, electri¬cal shock, and the resulting injuries will be discussed. The various electrical hazards will be described. Practices that will help keep you safe and free of injury are emphasized. To give you an idea of the hazards caused by electricity, case studies about real-life deaths will be described.

**PUB222**

**Excavator and Backhoe Safety – NIOSH**

Workers who operate or work near hydraulic excavators and backhoe loaders are at risk of being struck by the machine or its components or by excavator buckets that detach from the excavator stick. NIOSH recommends that injuries and death's be prevented through training, proper installation and maintenance, work practices, and personal protective equipment.

**PUB223**

**Exposure to Respirable Silica – NIOSH**

Occupational exposures to respirable crystalline silica occur in a variety of industries and occupations because of its extremely common natural occurrence and the wide uses of materials and products that contain it. At least 1.7 million U.S. workers are potentially exposed to respirable crystalline silica [NIOSH 1991], and many are exposed to concentrations that exceed limits defined by current regulations and standards. Occupational exposures to respirable crystalline silica are associated with the development of silicosis, lung cancer, pulmonary tuberculosis, and airways diseases.

**PUB224**

**Fatal Injuries to Workers –NIOSH**

It can happen to you. On average, 16 workers die each day in this country. These workers die simply trying to earn a living. This document includes 16 years of data from the National Traumatic Occupational Fatalities surveillance system for the years 1980 through 1995. Occupational injury mortality statistics on over 93,000 deaths are provided by demographic and injury characteristics. These data illuminate the nature and magnitude of work-related injury death for the United States and comprise the most comprehensive summary available in one document. Although fatal occupational injuries have decreased over the years, the burden remains high.

**PUB225**

**Health Effects of Worker Exposure to Asphalt – NIOSH**

This publication is an evaluation of the health effects of occupational exposure to asphalt and asphalt fumes. It includes an assessment of chemistry, health, and exposure data from studies in animals and humans exposed to raw asphalt, paving and roofing asphalt fume condensates, and asphalt-based paints. NIOSH recommends minimizing possible acute or chronic health effects from exposure to asphalt, asphalt fumes and vapors, and asphalt-based paints by preventing dermal exposure, keeping the application temperature of heated asphalt as low as possible, using engineering controls and good work practices at all work sites to minimize worker exposure to asphalt fumes and asphalt-based paint aerosols and to use appropriate respiratory protection. Increase worker awareness – share this publication.


**PUB226**

**Injuries and Deaths from Skid-Steer Loaders – NIOSH**

WARNING! Workers who operate or work near skid-steer loaders may be crushed or caught by the machine or its parts. The National Institute for Occupational Safety and Health (NIOSH) requests assistance in preventing injuries and deaths among workers who operate, service, or work near skid-steer loaders. NIOSH studies suggest that employers, supervisors, and workers may not fully appreciate the potential hazards associated with operating or working near skid-steer loaders and they may not follow safe work procedures for controlling these hazards. This Alert describes six deaths involving skid-steer loaders and recommends methods for preventing similar incidents.

## PUB227

### Organization of Work and Safety of Work – NIOSH

Revolutionary changes in the organization of work have far outpaced our knowledge about the implications of these changes for safety and health on the job. Research and development needs identified in this report include (1) improved surveillance mechanisms to better track how the organization of work is changing, (2) accelerated research on safety and health implications of the changing organization of work, (3) increased research focus on organizational interventions to protect safety and health, and (4) steps to formalize and nurture organization of work as a distinctive field in occupational safety and health.

## PUB228

### Pocket Guide to Chemical Hazards – NIOSH

The NIOSH Pocket Guide to Chemical Hazards provides a concise source of general industrial hygiene information. The Pocket Guide presents key information and data in abbreviated tabular form for 677 chemicals or substance groupings commonly found in the work environment (e.g., manganese compounds, tellurium compounds, inorganic tin compounds, etc.). The industrial hygiene information found in the Pocket Guide assists users to recognize and control occupational chemical hazards. The chemicals or substances contained in this revision include all substances for which the National Institute for Occupational Safety and Health (NIOSH) has recommended exposure limits (RELs) and those with permissible exposure limits (PELs) as found in the Occupational Safety and Health Administration (OSHA) Occupational Safety and Health Standards (29 CFR 1910.1000 – 1052).

## PUB229

### Practical Demonstrations of Ergonomic Principles – NIOSH

Workers don't have to have backaches or headaches at the end of the day. Train workers in the importance of considering ergonomic principles while working. This document was developed for individuals who intend to provide training on ergonomic principles that focus on risk-factor exposures of Musculoskeletal disorders (MSDs). It was designed for trainers of all experience levels including the beginning trainer. The demonstrations are designed to be performed by both the trainer and the worker. Each demonstration reinforces specific ergonomic principles and teaches the worker how and why to avoid MSD risk factors. Additionally, individuals involved in the purchase and selection of new and/or replacement tools may benefit from many of the demonstrations because they highlight the importance of considering ergonomic principles before purchasing tools.

## PUB230

### Preventing Falls from Communication Towers – NIOSH

Falls are deadly! Estimates vary greatly about the number of workers in telecommunication tower construction and maintenance. In 1993, estimates ranged from 2,300 to 23,000 workers in this field. These estimates suggest fatality rates of 49 to 468 deaths per 100,000 workers—nearly 10 to 100 times the average rate of 5 deaths per 100,000 workers across all industries. These deaths included 93 falls, 18 telecommunication tower collapses, and 4 electrocutions. However, the number of deaths identified here should be considered a minimum because identification methods are not exact.

## PUB231

### Preventing Worker Deaths in Trenches – NIOSH

There is no reliable warning when a trench fails. The walls can collapse sud¬denly, and workers will not have time to move out of the way. From 2000–2009, 350 workers died in trenching or excavation cave-ins—an av¬erage of 35 fatalities per year Factors such as type of soil, water con¬tent of soil, environmental conditions, proximity to previously backfilled ex¬cavations, weight of heavy equipment or tools, and vibrations from machines and motor vehicles can greatly affect soil.

## PUB232

### Prevention Through Design – NIOSH

This is one of those, "pay now, or pay later" concepts, and as in most situations, "pay now" costs less. Designing out hazards is the most effective means of preventing occupational injuries, illnesses, and fatalities. Although this concept is well known, there has not been a concerted effort to achieve broad implementation of it. In 2007, NIOSH initi¬ated a national initiative, Prevention through Design (PtD), to foster designing out occupational hazards in equipment, structures, materials, and processes that affect workers. Drawing on the knowledge of many different stakeholders, NIOSH is now presenting the Prevention Through Design Plan for the National Initiative. Consider incorporating into your work.

## PUB233

### Strategies for Improving Worker Training – NIOSH

Use this publication to enhance your training program. Although originally developed for training workers in the mining industry, this information could also apply to training workers in construction. This publication documents information presented in a series of workshops held during 2002 and 2003. Safety and health professionals have long recognized training as a critical element of an effective safety and health program, and a growing concern for safety professionals is the training of new workers.

## PUB234

### Teaching Young Workers Job Safety – NIOSH

Every year, approximately 53,000 youth are injured on the job seriously enough to seek emergency room treatment. This publication is the culmination of many years' work, dedicated to reducing occupational injuries and illnesses among youth. The activities in the Youth @ Work curriculum were developed in consultation with numerous teachers and staff from general high schools, school to work, work experience, and vocational education programs, as well as the California Work Ability program, which serves students with cognitive and learning disabilities. The activities have been used by numerous high school teachers, job trainers, and work coordinators around the country to teach youth important basic occupational safety and health skills.

## PUB235

### Worker Deaths by Falls – NIOSH

Over 80 occupational falls resulting in fatalities are reviewed. Very good real-world material for a safety meeting. Many workers, regardless of industry or occupation, are exposed to fall hazards daily. This publication describes the magnitude of the problem of occupational falls in the U.S., identifies potential risk factors for fatal injury, and provides recommendations for developing effective safety programs to reduce the risk of fatal falls. Based on the surveillance data, falls from elevations were the fourth leading cause of occupational fatalities from 1980 through 1994. The 8,102 deaths due to falls from elevations accounted for 10% of all fatalities and an average of 540 deaths per year.

## PUB236

### Audit for Aggregate Mining at Surface Metal/nonmetal – MSHA

This safety audit is focused on the most common violations found at aggregate operations. Twenty condition/practices accounted for a majority of all violations cited at sand, gravel and crushed stone mining operations in 2004.

## PUB237

### Guide for Safety Training, Plant Operations – MSHA

This Instructor Guide was developed to assist the sand, gravel, and crushed stone industry in conducting effective on-the-job training (OJT) of new employees, or employees reassigned to different jobs. This Instruction Guide uses a generic Job Safety Analysis (JSA) of jobs common to the industry. The JSA format facilitates uniform basic training in safe job procedures, while requiring only a minimum of time and effort on the part of the trainer.

## PUB238

### Machine Guarding – MSHA

Center for Disease Control (CDC) estimates close to 20,000 amputations in the workplace each year. Machine and equipment guarding help to prevent these types of injuries. This MSHA publication reviews guarding of conveyors and associated equipment. Although directed at mine safety, the mechanics of guarding is very appropriate for many other similar work situations. Workers need to understand the types of guards and systems at the workplace.

## PUB239

### Toolbox Training for Construction at Mines – MSHA

This Toolbox Training for Construction at Aggregate Mines contains 52 different modules designed to stimulate safety discussions among workers. Each module is intended to be completed in roughly 10 or 15 minutes, so a Toolbox Training session can be conducted once a week for the entire year. Every module follows the same format, so that as the weeks pass, workers will start to approach every safety situation in the same way. Some of the training topics include: working around equipment, PPE, materials handling, and hand tools. A blank format is included in the package to help develop your own topics.

## PUB240

### Assessment of Driver Drowsiness – FMCSA

One of the major findings of this study was evidence of a strong association between drowsiness and time of day. The early-morning time period between 6 a.m. and 9 a.m. was especially problematic for the Long and Short-Haul drivers. In this publication, researchers conducted a study to characterize episodes of driver fatigue and drowsiness and to assess the impact of driver drowsiness on driving performance.

## PUB241

### Large Truck and Bus Crash Facts 2009 – FMCSA

This annual edition of Large Truck and Bus Crash Facts contains descriptive statistics about fatal, injury, and property damage only crashes involving large trucks and buses in 2009. Selected crash statistics on passenger vehicles are also presented for comparison purposes. Great information to share with your drivers.

## PUB242

### Large Truck Crash Overview 2009 – FMCSA

The top two driver-related factors for large trucks and passenger vehicles in fatal crashes were the same: driving too fast (7% for trucks, 19% for passenger vehicles) and failure to keep in proper lane (6% and 18%). Of the 33,808 people killed in motor vehicle crashes in 2009, 10% (3,380) died in crashes that involved a large truck. Another 74,000 people were injured in crashes involving large trucks. Only 15% of those killed and 22% of those injured were occupants of large trucks.

## PUB243

### Understanding the FMCSA Cargo Securement Rules – FMCSA

Moving heavy vehicles, equipment and machinery? Or maybe concrete pipe, metal coils, logs or large boulders? This publication discusses the new cargo securement rules enacted by the Federal Motor Carrier Safety Administration (FMCSA). Knowing how many tie downs and where do you put them helps prevent articles from shifting on or within, or falling from commercial motor vehicles. Good publication – should be with every driver that operates cargo-carrying commercial motor vehicles.


## PUB244

### Weather Impact on Large Truck Safety – FMCSA

This paper examines different types of weather events as safety risk factors in Commercial Motor Vehicle (CMV) crashes, including the possible role of climate change in altering the distribution, frequency, or severity of those weather events. CMVs are different from other vehicles with respect to weather and weather-related crashes. They have greater mass and thus a greater release of kinetic energy during a crash. Share this publication with your drivers – this will get them talking!


## PUB245

### Manual of Uniform Traffic Control Devices - FHWA

The purpose of traffic control devices, as well as the principles for their use, is to promote highway safety and efficiency by providing for the orderly movement of all road users on streets, highways, bikeways, and private roads open to public travel throughout the Nation. Traffic control devices notify road users of regulations and provide warning and guidance needed for the uniform and efficient operation of all elements of the traffic stream in a manner intended to minimize the occurrences of crashes.

## PUB246

### Work Zone Operations, Best Practices – FHWA

This Work Zone Best Practices Guidebook provides an easily accessible compilation of work zone operations practices used and recommended by various States and localities around the country. The best practices are grouped into 11 major categories to help practitioners easily find practices that deal with a particular topic. Practices can also be found via 7 cross-references that enable users to find best practices in several different ways, and a subject index that offers 50 topics and subtopics for more specific searches.

## PUB247

### California Manual on Uniform Traffic Control Devices. Temporary Traffic Control (Parts 1 & 6). – Caltrans/FHWA

This Manual contains the basic principles that govern the design and use of traffic control devices for all streets, highways, and bikeways, regardless of type or class or the public agency, official, or owner having jurisdiction. It specifies the restriction on the use of a device if it is intended for limited application or for a specific system.

## PUB248

### California Occupational Safety and Health Program – Cal/OSHA

This publication explains the requirements of California law for workplace safety and health, and the functions of the California Occupational Safety and Health (Cal/OSHA) Program.

## PUB249

### Employer Sample Procedures for Heat Illness Prevention (template) – Cal/OSHA

California Employers with any outdoor places of employment must comply with the Heat Illness Prevention Standard T8 CCR 3395. These procedures have been created to assist the employer in crafting their heat illness prevention procedures, and to reduce the risk of work related heat illnesses among their employees.

**PUB250**

## Ergonomic Guidelines for Manual Material Handling – Cal/OSHA

Manual material handling (MMH) work contributes to a large percentage of the over half a million cases of musculoskeletal disorders reported annually in the United States. Musculoskeletal disorders often involve strains and sprains to the lower back, shoulders, and upper limbs. This booklet will help you to recognize high-risk MMH work tasks and choose effective options for reducing their physical demands.

**PUB251**

## Ergonomic Survival guide for Carpenters and Framers – Cal/OSHA

This Survival Guide is designed to promote awareness of safe work practices for Carpenters and Framers.

**PUB252**

## Ergonomic Survival Guide for Cement Masons – Cal/OSHA

This Survival Guide is designed to promote awareness of safe work practices for Cement Masons.

**PUB253**

## Ergonomic Survival Guide for Electricians – Cal/OSHA

This Survival Guide is designed to promote awareness of safe work practices for Electricians.

**PUB254**

## Ergonomic Survival Guide for Laborers – Cal/OSHA

This Survival Guide is designed to promote awareness of safe work practices for Laborers.

## PUB255

## Ergonomic Survival Guide for Sheet Metal Workers – Cal/OSHA

This Survival Guide is designed to promote awareness of safe work practices for Sheet Metal Workers.

## PUB256

## Farm Labor Contractor Safety and Health Guide – Cal/OSHA

The farm labor contractor, like all other employers, holds ultimate responsibility for the health and safety of his or her employees. Operating in full compliance with the law is challenging but not impossible. This guide was written to help you protect your workers and to prevent or reduce the high number of injuries and illnesses that occur in the agricultural industry.

## PUB257

## Guide to California Hazard Communication Regulation – Cal/OSHA

Every day at workplaces throughout California, employees work with or are incidentally exposed to hazardous substances that can harm their health or cause safety hazards. This guide is designed to help employers and employees understand the requirements of the hazard communication regulation by providing a simplified and clear overview of the major program elements. For easy reference, this guide is separated into seven main sections:

- Scope
- Hazard Determination
- Material Safety Data Sheets (MSDSs)
- Labels and Other Forms of Warning
- Written Hazard Communication Program
- Employee Information and Training
- Trade Secret Protection

## PUB258

### Guide for Working Safely With Supported Scaffolds – Cal/OSHA

This GUIDE promotes awareness of safe work practices for supported scaffolds and covers:

- Commonly Used Supported Scaffolds
- Common Hazards
- Selected CAL/OSHA Regulations
- Working Safely – Best Practices

## PUB259

### Is It Safe To Enter A Confined Space? – Cal/OSHA

For easy reference, the guide is separated into six distinct main sections:

1. Rescue
2. Definitions and Basics
3. Confined Space Hazards
4. Hazards Control
5. Training and Education
6. Frequently Asked Questions

## PUB260

### Keys to Success and Safety for the Construction Foreman – Cal/OSHA

An Ergonomic Approach to Cost Reduction

## PUB261

### Lead In Construction – Cal/OSHA

Cal/OSHA is conducting a Special Emphasis Program to reduce the hazard from lead in construction affecting workers, their families and the public.

## PUB262

### Protect Yourself from Heat Illness – Cal/OSHA

Heat illness includes heat cramps, fainting, heat exhaustion, and heatstroke. Workers have died or suffered serious health problems from these conditions. Heat illness can be prevented. Watch for symptoms in yourself and your coworkers. If you feel any symptoms, tell your coworkers and supervisor immediately because you may need medical help. Know who to talk to and how to get help before you start each workday.

## PUB263

### Respirator Regulation – Cal/OSHA

Cal/OSHA's regulation for worker use of respirators is Section 5144 in Title 8, California Code of Regulations. Section 5144 details steps employers must take to assure safe and effective use of respirators in the workplace.

## PUB264

### Safety and Health Training And Instruction Requirements – Cal/OSHA

The following is a list of the instruction and training requirements contained in the Construction Safety orders (Subchapter 4) and the General Industry Safety Orders (Subchapter 7) of Title 8, Division 1, Chapter 4 (with several references contained in Chapter 3.2) of the California Code of Regulations. Also included are references to both Competent Person and Qualified Person.

## PUB265

### Silica Hazard Alert – Cal/OSHA

Exposures to respirable crystalline silica dust during construction activities can cause serious respiratory disease. Each year more than 300 U.S. workers die from silicosis and thousands more are diagnosed with the lung disease.

**PUB266**

**Workplace Injury and Illness Prevention Program for High Hazard Employers (template) – Cal/OSHA**

This model program has been prepared for use by employers in industries, which have been determined by Cal/OSHA to be high hazard. The requirements for establishing, implementing and maintaining an effective written injury and illness prevention program consist of the following eight elements:

1. Responsibility
2. Compliance
3. Communication
4. Hazard Assessment
5. Accident/Exposure Investigation
6. Hazard Correction
7. Training and Instruction
8. Recordkeeping

**PUB267**

**Guide for Cranes and Derricks in Construction – OSHA**

This guide is intended to help businesses comply with OSHA's standard for Cranes and Derricks in Construction. It is designed to address the most common compliance issues that employers will face and to provide sufficient detail to serve as a useful compliance guide. It does not, however, describe all provisions of the standard and the reader must note that California (Cal/OSHA) has several additional, more stringent, rules.

# List of Acronyms

ACCM: asbestos-containing construction material

ACGIH: American Conference of Industrial Hygienists

ACM: asbestos-containing material

AEGC program: assured equipment grounding conductor program

ANSI: American National Standards Institute

ASSE: American Society of Safety Engineers

ASTM: American Society for Testing and Materials

ATSSA: American Traffic Safety Services Association

°C: Degree Celsius temperature scale

Cal/OSHA: California Occupational Safety and Health Administration

Ca PE: California Registered Professional Engineer

CARB: California Air Resources Board

CASOs: Compressed Air Safety Orders

CAZ: controlled access zone

CCR: California Code of Regulations

CFR: Code of Federal Regulations

CO2: carbon dioxide

CSHIP: Construction Safety and Health Inspection Project

CSOs: Construction Safety Orders

cu.ft.: cubic feet

cu.yd.: cubic yard

d: Penny size of nails

dBA: a unit of sound level as measured on the A-scale of a standard sound level meter

DOSH: Division of Occupational Safety and Health

EMS: emergency medical service

ESOs: Electrical Safety Orders

eTool: electronic educational products for safety and health

°F: Degree Fahrenheit temperature scale

FP: fall protection

FPP: fall protection plan

ft.: feet

GFCI: ground-fault circuit interrupter

GISOs: General Industry Safety Orders

haz-com program: hazard communication program

HEPA: high-efficiency particulate air

HP: hearing protection

IDLH: immediately dangerous to life or health

IIP Program: Injury and Illness Prevention Program

in.: inches

ISEA: International Safety Equipment Association

LAZ: limited access zone

LEL: lower explosive limit

MSDS: material safety data sheet

MSHA: Mine Safety and Health Administration

NFPA: National Fire Protection Association

NIOSH: National Institute for Occupational Safety and Health

NOx: Oxides of Nitrogen

o.c.: On center

OPU: order to prohibit use

PACM: presumed asbestos-containing material

PAT: powder-actuated tool

PEL: permissible exposure limit

PFA: personal fall arrest

PFP: personal fall protection

PFR: personal fall restraint

PPE: personal protective equipment

psf: pounds per square foot, unit of pressure

psi: pounds per square inch, unit of pressure

QP: qualified person

RMI: repetitive motion injury

SAR: supplied-air respirators

SO: safety order

sq.ft.: square feet

T8 CCR: Title 8 of the California Code of Regulations

tsf: tons per square foot

TSOs: Tunnel Safety Orders

TWA: time-weighted average

V: volt, unit of electric voltage

# Index

---

**Qualified Person, 122**

railings, 52, 73 ,82, 83, 124, 125, 126, 136, 141, 145, 148

**Ramps and Runways, 122-123**

    foot ramps, 123,124

    wheelbarrow ramps, 123

rebar, 22, 24, 26, 69

    fall protection, 69

    impalement protection, 24, 25

    supporting of rebar, 27

recordkeeping, 4 6, 98

registration, 4, 5, 6, 8, 14, 99

repetitive motion injury, 54, 55

reporting, 6, 18, 20, 55, 86, 99

    accidents, 6

    blasting, 18

    carcinogens, 20

    heat illness, 86

    IIPP, 99

    RMIs, 55

respiratory protection, 4, 16, 27, 29, 41, 95, 117, 118, 165

    asbestos, 16

    confined spaces, 27,29

    dust, 41

    program, 4

    welding, 165

rollover protection, 78, 94

**Roofing Operations, 123-128**

    fall protection methods, 123, 124, 125, 126

    fall protection requirements, 127, 128

    hot operations, 127, 128

    multi-unit roof, 126

    new-production type, 124, 126, 127

    roof openings, 127, 128

    single unit roof, 124, 125, 127

safety

    apparel, 76, 93, 158, 159

    conference/meeting, 8, 21, 98, 161, 162

    factor, 68,69

    harness, 29,140

    monitoring system, 72

    nets, 67,69

    precautions, 21,98

    programs, 8, 98, 100

    training, 98,106,110

    valve, 12

    workplace, 1, 99, 161

sanitation, 121, 149

saws, 17, 152, 155, 156-157

    band saw, 156

    chain saw, 152, 157

    circular saw, 156

    guard, 155

    miter saw, 155

    radial arm saw, 156

    speed, 155, 156, 157

    table saw, 156

**Scaffolds, 129-145**

    access, 130, 139

    design and construction, 129, 130

    erecting and dismantling, 130

    general requirements, 129-135

    height limits, 136-137

    horse scaffold, 134, 144

    ladder jack scaffold, 144

    planking, 132, 133, 134

## Ordering and Contact Information

Mike Leuck coaches Construction Professionals in safety compliance and delivers a fast paced, compliance targeted presentation to Supervisors.

To schedule Mike, see "Supervisor Training."

www.oshatools.com

Mike built the "Cal/OSHA Compliance Guide" from his years of walking the job with Construction Supervisors and reviewing hundreds of Agency publications. Every Supervisor should have a copy in their toolbox.

Place your order online

www.oshatools.com.

Mike Leuck

Tailgate Publications

mikeleuck.csp@gmail.com

(831) 212 0093

## About Mike Leuck

Mike's passion is guiding Construction Supervisors in job site safety compliance. He delivers a fast paced, compliance targeted presentation at Supervisor meetings and is available for site visits.

He is experienced in high-hazard industries, including construction, marine, transportation, mining, aviation and telecommunications. He holds certifications from the Board of Certified Safety Professionals (CSP/STS) and the International Society of Mine Safety Professionals (CMSP). He is an authorized Trainer for both the OSHA Construction and General Industry 10/30 Hour course.

- Certified Safety Professional (CSP)
- Certified Mine Safety Professional (CMSP)
- Safety Trained Supervisor (STS)
- OSHA Instructor – Construction and General Industry

Mike built the "Cal/OSHA Compliance Guide" from his years of walking the job with Construction Supervisors and reviewing hundreds of Agency publications. Every Supervisor should have a copy in their toolbox.

www.ingramcontent.com/pod-product-compliance
Lightning Source LLC
Chambersburg PA
CBHW061213220326
41599CB00025B/4624